NAVY & MARINE CORPS FIRE APPARATUS 1836 - 2000

William D. Killen

Iconografix

Photo Gallery Series

Iconografix
PO Box 446
Hudson, Wisconsin 54016 USA

Library of Congress Card Number: 00-132300

ISBN 1-58388-031-3

00 01 02 03 04 05 06 5 4 3 2 1

Printed in the United States of America

Cover and book design by Shawn Glidden

Edited by Dylan Frautschi

COVER CREDIT: Naval Air Station Oceana's 1998 Emergency One Cyclone TC photographed by Warren Liebmann, www.silverreflection.com

BACK COVER:
Boston Navy Yard's 1954 American LaFrance 700 Series open cab pumper was one of two pumpers operated at the Chelsea conflagration in 1973 by Boston Navy Yard firefighters Jack Carpenter, Kevin Glynn, and Harry Tagen. Carpenter retired as Fire Chief, Brunswick Naval Air Station, Glynn retired as Fire Chief, South Weymouth Naval Air Station, and Harry Tagen is currently the Fire Chief at the Portsmouth Naval Shipyard, Portsmouth, New Hampshire.

BOOK PROPOSALS

Iconografix is a publishing company specializing in books for transportation enthusiasts. We publish in a number of different areas, including Automobiles, Auto Racing, Buses, Construction Equipment, Emergency Equipment, Farming Equipment, Railroads & Trucks. The Iconografix imprint is constantly growing and expanding into new subject areas.

Authors, editors, and knowledgeable enthusiasts in the field of transportation history are invited to contact the Editorial Department at Iconografix, Inc., PO Box 446, Hudson, WI 54016.

ACKNOWLEDGMENTS

My sincere thanks to everyone who assisted me in completing this pictorial history of Navy and Marine Corps fire apparatus. This book was made possible through the collective efforts of so many friends and fire buffs throughout the United States and overseas.

The following personnel were of special help with data, photographs, slides and interviews:

Richard Adelman
Jim Atkinson
Roger Bjorge
John M. Calderone
Stuart Cook
Crash Rescue Equipment Corporation
Raoul Denton
Jim Derstine
Warren Dixon
Emergency One, Inc.
Fire Apparatus Journal
Victor Flint
FWD-Seagrave Corporation
Tom Gemp
Chris Hatch
Doug Howard
J. R. Hunneman
Lynn Johnson
Buddy King
Kevin King
Kovatch Mobile Equipment
Warren Liebmann

Joe MacDonald
James Mansbery
James Manser
Danny Miller
John Peckham
Ed Peterson
Mark Redmon
Arthur Richardson
John Rieth
Carol Shelton
Chris J. Scheer
Tom W. Shand
Dennis C. Sharpe
Harry Tagen
Carl Thomann
Doug Thomas
John J. Wentzel
Peter West
Jaimie Wood
Joel C. Woods
Larry W. Wools

Special thanks to Don Curtis, Navy Seabee Logistics Center, Port Hueneme, California, June Heninger, Navy Transportation Equipment Management Center, Norfolk, Virgina, and Gary Lind, Director of Transportation, U.S. Navy, for their technical assistance.

Many of the photographs and slides in my collection are not identified with the name of the person who took the picture. Those photographs are identified as part of my collection since I could not correctly credit the photographer. I apologize to any contributor whose name is not listed above.

Last, but certainly not least, my deepest appreciation to my best friend and wife of 41 years, Carole, for her assistance and support in completing this project.

INTRODUCTION

This book is the result of several conversations with my good friend, Ed Peterson, President of the Society for the Preservation and Appreciation of Antique Motorized Fire Apparatus in America. Last October Ed and I strolled through the Green and Chocolate Fields at the Antique Automobile Club of America's Eastern Regional Fall Meet browsing and discussing Ed's soon to be published book on Fire Chiefs cars. I shared with Ed my desire to publish the history of Stoughton Wagon Works, the manufacturer of my 1923 Stoughton fire engine. As our stroll and discussion continued, I got the idea to put together a book on Navy and Marine Corps fire apparatus.

Having collected photographs, slides, operator's manuals, and advertisements of fire apparatus for more than thirty years, I decided to explore the idea to publish a book on Navy fire apparatus. Since I have an extensive collection of photographs, particularly military apparatus, and many of my friends and fire buffs provided me with copies of the photographs and slides they took of Navy fire apparatus over the years, I assembled a pictorial history of Navy and Marine Corps fire apparatus.

What began as a history spanning the period 1896 to 1999, soon changed as my research led me to more information about Navy fire apparatus at the U.S. Naval Academy in 1866. Ed Tufts' book, "Hunneman's Amazing Fire Engines" led me to Chief Arthur Richardson of the Boothbay Harbor, Maine, fire department and J. R. Hunneman, the great-great-great-grandson of William C. Hunneman, founder of the Hunneman Company. The Hunneman Company delivered a hand engine to the Boston Navy Yard in 1836, then known as the Charlestown Navy Yard for the sum of $700.00. This engine was later sold to Boothbay Harbor, Maine, where it was renamed the "Minnehaha," where it still resides today.

The extent of documented Navy fire service history is about a page and a half, published in OPNAV-P415-106 January 15, 1947, which reads: "In the not too distant past, the Navy's firefighting program ashore consisted mainly of volunteer fire brigades. Hose-reel carts and soda-acid chemical wagons were spotted around the activity at various locations, and when a fire broke out, the bugle sounded fire call, the ship's bell was rung, and everyone dropped what he was doing and dashed off to see where the fire was. Men were assigned certain fire stations in advance on the watch quarter and station bill, and even if the station happened to be a remote fire barrel and bucket, the assigned men would run clear across the compound to stand by it, regardless of where the fire turned out to be. Under such a plan, the activity of the entire Naval Establishment was disrupted for every smoke scare or minor blaze."

"With the advent of the Second World War and the acquisition of millions of dollars in property values on Naval Shore Establishments, it was readily perceived that fire protection should be provided on a more certain and efficient basis. With public fire departments overtaxed and undermanned, it was found impracticable in most instances to place entire reliance on such outside aid. It did not take many large fires at the outset of the war to confirm the realization that full-time fire protection was a necessity."

"Owing to the increasing demand for the utilization of every able-bodied man afloat, it was soon found desirable at shore installations to replace military personnel with civilian firefighters, as far as the latter were available under the prevailing critical manpower shortage. The majority of the civilians who were subsequently hired as a result of this policy were retired municipal firemen, who despite their advanced years were able to provide the supervision, training, and leadership for either the civilian or military personnel (or both) who made up the rosters of naval fire departments. Prior to the surrender of Japan, fire departments on naval activities were becoming well organized, equipped, manned and trained."

This pictorial history of Navy and Marine Corps fire apparatus may serve to provide a little more information on the history of the Navy and Marine Corps fire services and is dedicated to those men and women, both past and present, who serve in protecting the naval installations and dependents of the sailors and marines who defend our Country.

The oldest known surviving U.S. Navy fire engine is Hunneman number 168 built for the Charlestown Navy Yard, Boston, Massachusetts. This engine was delivered to the Navy on March 7, 1836 for $700.00, and was later sold to Boothbay Harbor, Maine, where it was renamed "Minnehaha." The Navy Yard opened in 1800 and may have purchased fire engines from other manufacturers. CREDIT: Chief Arthur Richardson

Boothbay Harbor firefighters give "Minnehaha No. 1" a cleaning and polishing in preparation for a muster and parade. This engine was on display at the Lincoln County Fire Museum for a number of years and now resides at the Boothbay Harbor Fire Station, Boothbay Harbor, Maine. CREDIT: Chief Arthur Richardson

The Franklin Pierce is a 7-inch Hunneman, delivered to the Navy Yard, Portsmouth, New Hampshire, on September 23, 1854. From Portsmouth it was sent to the Navy Yard at Norfolk, Virginia, and later was taken to the Government Yard at Pensacola, Florida. From Pensacola it was returned to Portsmouth, where it was condemned. The Portsmouth Volunteer Fireman's Association purchased the Hunneman Engine from the Navy and leased it to the Franklin Pierce Volunteer Fireman's Association. The engine was scrapped in 1948.

This 1866 Horse Drawn Double, Straight Frame Second Size Amoskeag Steamer, known as the "Severn," was delivered in October 1866 and was the U.S. Naval Academy's first steam engine. This photograph was taken circa 1899. CREDIT: Doug Thomas, Fire Marshal Administrator, Retired.

"Balance hose cart sold to Charlestown Navy Yard" per note by John C. Hunneman on reverse of the original photograph. The "balance" means the toolbox was on the stern to help balance the weight of the drawbar. A. H. Folsom took this photo outside one of the Hunneman factory buildings about 1868-1878. CREDIT: J. R. Hunneman, Jr.

This 1875 Ryan Brothers hose carriage was the third hand-drawn hose carriage delivered to the U.S. Naval Academy. The first was delivered in 1849 and the second was delivered in 1859. It is possible the Hunneman Company made these hose carriages, since Hunneman Engines 361 and 634 were delivered in 1849 and 1859 respectively. The hose carriage is incorrectly identified as "1879" in this photograph. CREDIT: Doug Howard

The U.S. Naval Academy personnel pose with the Academy's 1866 Amoskeag steam engine and 1875 Ryan Brothers hose carriage during a training evolution. Circa 1885. Credit: U.S. Navy

The U.S. Naval Academy placed this "American" Extra First Size steam fire engine in service in 1896. This engine was built by the American Fire Engine Company, registry number 2457, and is unique in that it is rigged for draft by hand or horse and is peculiar to the Naval Academy due to the presence of so many midshipmen seen here pulling the American. CREDIT: U.S. Navy

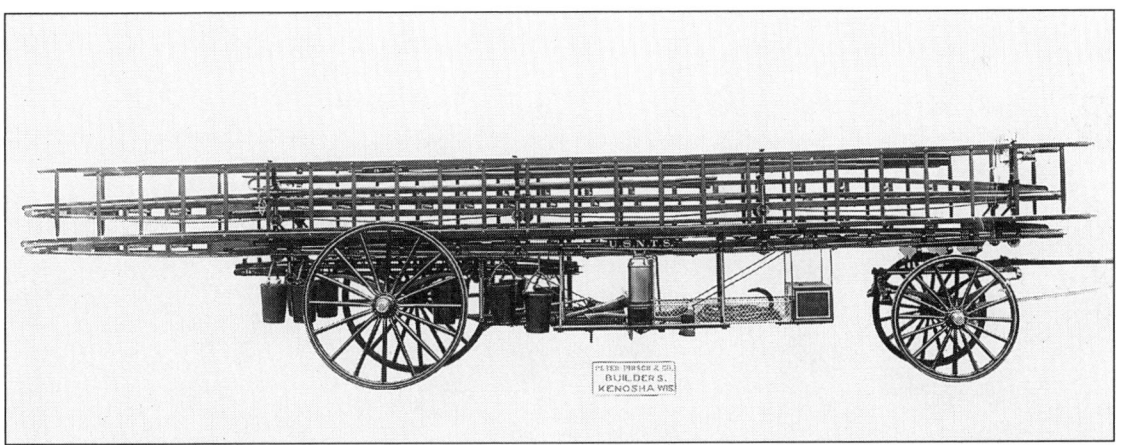

The Naval Training Station where this 1900 era horse-drawn Peter Pirsch ladder was assigned is unknown. In addition to a complement of ladders, the rig carried soda acid fire extinguishers, fire buckets, a Peter Pirsch hose clamp, axes, and a tool box. The letters "U.S.N.T.S" are most likely "U.S. Naval Training Station" and could have been NTS Newport, Rhode Island (opened 1881), and NTS Yerba Buena Island, San Francisco, California (1889). It is possible NTS Norfolk, Virginia (1908), or NTS Great Lakes, Illinois (1911), may have used this ladder. CREDIT: Bill Killen Collection

The year this horse-drawn American LaFrance "Metropolitan" steam fire engine was manufactured and the Navy base where it served are unknown. The letters "USN" can be seen on the tank below the driver's seat. CREDIT: Jim Atkinson Collection

Portsmouth Naval Shipyard, Portsmouth, New Hampshire, used this steam fire engine until the 1930s. The steamer has "Ashland" on the lantern and on the coal box at the rear of the boiler. CREDIT: Portsmouth Naval Shipyard Fire Department

Naval Station Newport, Rhode Island, used this 1916 Dodge Touring car as a Fire Chief buggy. The fire trucks are an American LaFrance Type 10, a Seagrave 750-gpm pumper, and a Southbend pumper. CREDIT: Jim Atkinson Collection

This 1917 American LaFrance Type 75 pumper, registration number 1939, was shipped to the Navy in the Panama Canal Zone on December 28, 1917. CREDIT: Bill Killen Collection

Richard Adelman obtained this photograph of an Ahrens Fox Model K4 from the archives of the Smithsonian Institution, Washington, DC. The location of the Navy base where this photo was taken is unknown. CREDIT: Richard Adelman Collection

This American LaFrance Type 40 Combination pumper was shipped to Marine Corps Recruit Depot, Parris Island, South Carolina, on November 15, 1918. It is probable that the Marine Corps purchased two Type 40 models based on information that the registration number is either 2395 or 2396, implying that two engines were supplied. CREDIT: Bill Killen Collection

This 1919 Ford Model TT 1-ton, with Peter Pirsch Chemical apparatus added, is lettered "G.L.N. Ordnance Depot." The identity of the naval base where it was assigned is unknown. CREDIT: Roger Bjorge Collection

Rear view showing the chemical tanks, Dietz lanterns, and soda acid extinguishers on the 1919 Model TT 1-ton Peter Pirsch. CREDIT: Roger Bjorge Collection

This 1919 American LaFrance Type 75 Triple on a Ford Model T chassis, registration number 2766, was shipped to the Norfolk Navy Yard, Norfolk, Virginia, September 30, 1919. CREDIT: Bill Killen Collection

U.S. Naval Proving Ground, Dahlgren, Virginia's first fire truck (circa 1920) was built on this Ford Model TT chassis. The unit was equipped with a chemical tank and booster hose, Dietz lanterns, a Sterling hand crank siren, and two sections of hard suction hose. CREDIT: U.S. Navy Photo

This 1924 Model T Chemical car was delivered to the U.S. Naval Academy, Annapolis, Maryland, in 1924 by the Colonial Motor Company of Annapolis, Maryland. The manufacturer of the firefighting package is unknown. CREDIT: Doug Thomas, Retired Navy Fire Marshal Program Administrator

This 1926 Peter Pirsch Special Foam unit was built for the Navy Yard, Brooklyn, New York, and delivered June 10, 1926. The wheelbase was 178 inches and the drive train was a Wisconsin "Y" motor with a Brown & Lipe transmission and clutch. The Pirsch customer order number 66336 details the specifications and equipment ordered by the Navy. CREDIT: Bill Killen Collection

The Mare Island Navy Shipyard, Vallejo, California, ordered this 1927 Peter Pirsch combination fire truck July 12, 1927. The truck was equipped with a steel hose body with a capacity of 750 feet, two 10-foot lengths of 4 1/2-inch hard suction hose, a variety of adapters and appliances, and more than 200 feet of ground ladders. The engine was a Waukesha 6AB with a manual transmission and a 12-volt electrical system. The truck was painted Standard Navy Gray and had a nickel plate finish throughout. CREDIT: Bill Killen Collection

This photo of the PCTS Fire Station in Keyport, Washington, was taken in July 1927. The "TS" in "PCTS" probably is Torpedo Station. The year and manufacturer of the hose wagon is unknown. The Navy is well known for changing the name of Navy bases as missions or functions change. CREDIT: Bill Killen Collection

This is a rear view showing a double bank ladder rack on Mare Island Navy Yard's 1927 Pirsch quadruple combination. CREDIT: Roger Bjorge Collection

This 1929 GMC-American LaFrance was built for the U.S. Naval Ammunition Depot at Fort Mifflin, Pennsylvania. This special unit was equipped with midship mounted chemical tanks with hose reels mounted above. The carbon dioxide cylinders have lines that appear to connect to the chemical cylinders and may have been used to pressurize the chemical tanks. CREDIT: Jim Atkinson Collection

Portsmouth Naval Shipyard's fire department was manned with Marines prior to World War II. This photograph shows the Marine Corps crew with the "old hose wagon" and the steam fire engine in front of the old fire station in Building 64. CREDIT: Portsmouth Naval Shipyard Fire Department

Marines and sailors watch as Portsmouth Naval Shipyard firefighters supply hose lines from their steam fire engine. CREDIT: Portsmouth Naval Shipyard Fire Department

Portsmouth Naval Shipyard firefighters draft water with what is believed to be a Hunneman engine from a rowboat that is being filled by the steamer. CREDIT: Portsmouth Naval Shipyard Fire Department

Portsmouth Naval Shipyard firefighters with an unidentified hand pumper that is probably a Hunneman. The U.S. Navy purchased several Hunneman engines for ships and naval stations. CREDIT: Portsmouth Naval Shipyard Fire Department

The Navy base, year, manufacturer, and carbon dioxide fire extinguishing system of this motorcycle are unknown. Roy Wolley, Editor of Fire Engineering magazine sent this photo to Richard Adelman several years ago. CREDIT: Richard Adelman Collection

Portsmouth Naval Shipyard Fire Department's "Hook and Ladder Wagon," followed by their "Stutz pumping engine," were photographed in 1933. The ladder truck is circa 1930s and the Stutz is early 1920s and has hard rubber tires on wooden wheels. CREDIT: Portsmouth Naval Shipyard Fire Department

Barton Fire Pumps advertised their 1933 Barton American 500-gpm pumper delivered to the U.S. Naval Academy Dairy, Millersville, Maryland, in the November 1933 issue of Fire Engineering. CREDIT: Ed Bosanko Collection

P-1502

This unique fire truck was built for the Navy's Lualualei Ammunition Depot in Oahu, Territory of Hawaii. The 24-inch Budd wheels were specifically built similar to a railroad wheel and were fitted with Firestone 5.50 x 24 pneumatic tires and tubes. This front view shows the "railroad" style bumper and the wheel assembly. The truck was painted Hamilton Red. CREDIT: Bill Killen Collection

This 1934 Peter Pirsch Model 21, 500-gpm Rail-type fire engine was ordered by the U.S. Navy Department's Bureau of Supplies on November 17, 1933. The truck was equipped with a 500-gpm type "S" pump with bronze pump body and a 150-gallon water tank. CREDIT: Bill Killen Collection

This 1934 Seagrave 500-gpm pumper was assigned to a Navy base in San Francisco, California. This unit was equipped with a single-stage pump and two booster reels. CREDIT: Bill Killen Collection

This unusual fire apparatus built on a 1934 General Motors truck was probably built in a Navy Public Works maintenance shop. The type of pump and water tank capacity is unknown. CREDIT: Jim Atkinson Collection

The Peter Pirsch Company introduced a new restyled hood and V-type radiator in 1935. This 1935 Pirsch 500-gpm rotary gear pumper is equipped with a carbon dioxide extinguishing system and lettered "Naval Torpedo Station." According to factory records this unit was delivered to the Navy base in Keyport, Washington. CREDIT: Bill Killen Collection

This early 1936 Peter Pirsch 500-gpm Model 20 pumper was powered with a 6-cylinder, 125-hp Waukesha motor and was assigned to the Naval Station Keyport, Washington. CREDIT: Roger Bjorge Collection

The Washington Navy Yard fire department in Washington, DC, was pressed into service to pump water from Navy buildings during the March 19, 1936 flood. The American LaFrance and Ahrens Fox pumpers have solid rubber tires and are believed to be early 1920s or earlier. The unidentified rig in the center appears to be a hose wagon and is equipped with pneumatic tires. CREDIT: J. Chris Scheer Collection

Buffalo Fire Appliance Corporation advertised "Buffalo Better Built Fire Apparatus and Fire Extinguishers" in the Buffalo Times Rotogravure Section in 1937. The advertisement showed U.S. Marine Barracks, Quantico, Virginia's 1937 750-gpm two-stage centrifugal pumper. CREDIT: Peter West

The Navy purchased several Howe pumpers in the late 1930s. This factory photo (circa 1937) shows a typical Navy 500-gpm pumper. The chassis builder is unknown. CREDIT: Richard Adelman Collection

U.S. Naval Academy's 1937 American LaFrance Type 415 CDB 1500-gpm Quad bears serial number 7795. This V-12 powered engine developed 240 hp. CREDIT: Richard Adelman Collection.

Old Navy fire trucks are often found in odd places. This ex-U.S. Naval Ammunition Depot, Hingham, Massachusetts, 1938 Mack was discovered in Milford, Connecticut. CREDIT: Bill Killen Collection

This 1938 GMC chemical-carbon dioxide unit served the Puget Sound Naval Shipyard, Bremerton, Washington, until taken out of service in the early 1950s. Today this unit is used to promote Public Fire Safety Education, public relations events, and parades. CREDIT: Lynn Johnson

The Navy base where this 1938 White tandem-axle chassis crash truck was assigned is unknown. It is safe to assume it was assigned to a Naval Air Station because of the carbon dioxide firefighting system mounted behind the driver's seat. CREDIT: Jim Atkinson Collection

The U.S. Marine Corps used this 1939 cab over engine Chevrolet to provide airfield fire protection. The capacity of the carbon dioxide firefighting system, as well as the Marine Corps base where this unit was assigned is unknown. CREDIT: Jim Atkinson Collection

Norfolk Naval Station, Norfolk, Virginia used this 1939 Peter Pirsch Model 14, 1000-gpm pumper for 23 years. The Model 14 was powered by a Hercules Model HXE engine and was sold to Staunton, Virginia, on August 15, 1962. CREDIT: Bill Killen Collection

This 1939 Maxim 500-gpm pumper was built on an International D35 chassis by Maxim for the Portsmouth Naval Shipyard, Portsmouth, New Hampshire. It had a 200-gallon water tank and carried a large assortment of ladders. This truck was sold to the Redstone Volunteer Fire Company, Conway, New Hampshire. CREDIT: Don Mason Collection

This 1939 Maxim aerial ladder was assigned to Naval Air Station Miramar, San Diego, California. CREDIT: U.S. Navy

The Navy Advanced Base Depot in Davisville, Rhode Island, used this 1939 Maxim aerial ladder. CREDIT: Bill Killen Collection

This W.S. Darley-General Motors truck chassis was equipped with a 500-gpm pump and was delivered to the U.S. Naval Station Tutuila, Samoa, in 1940. CREDIT: Bill Killen Collection

This 1940 Buffalo 1000-gallon "Commander" triple combination served the Navy Yard, Philadelphia, Pennsylvania. CREDIT: Bill Killen Collection

This 1940 Series 185 Seagrave 750-gpm pumper was 1 of 12 V-12 powered pumpers delivered to the Navy. This unit was delivered to the Bureau of Aeronautics and may have been assigned to the Naval Air Station at Astatula, Florida. CREDIT: Bill Killen Collection

American LaFrance built several carbon dioxide fire apparatus such as this unit on a 1940 Ford COE chassis. CREDIT: Richard Adelman Collection

Brooklyn Navy Yard's 1940 Series 160 pumper was powered by a Straight 8 motor and delivered 500 gpm. The unit carried 1000 feet of 2 1/3-inch hose and 200 gallons of water. CREDIT: Bill Killen Collection

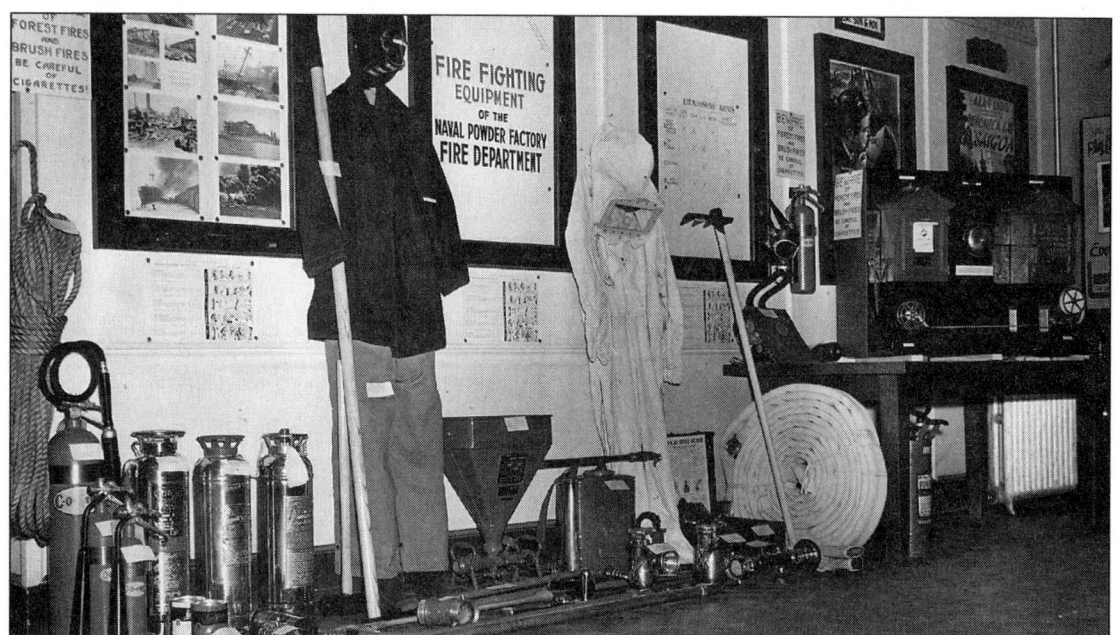

Fire alarm system components and firefighting equipment used by the Naval Powder Factory fire department, Indian Head, Maryland, is displayed during fire prevention week circa 1942. CREDIT: Bill Killen Collection

Portsmouth Naval Shipyard Fire Department's 1942 Ford "Chief's Beachwagon," 1940 Ford Chemical with 500 pounds of carbon dioxide and Mobile Firefighting Unit carried shipboard firefighting equipment and is believed to be an International Harvester 2 1/2-ton 6 x 6. CREDIT: Portsmouth Naval Shipyard Fire Department

Naval Powder Factory, Indian Head, Maryland's 1940 GMC pumper, 1933 Peter Pirsch pumper, and 1933 Chevrolet Coupe Fire Chief's car parked in front of Building 320, the second fire station built in 1919. CREDIT: Frank Cotrufo, Jr. Collection

The Long Island Fuel Depot was located on Long Island in Casco Bay, near the Casco Bay Destroyer Base, Casco Bay, Maine, which opened in 1940. American LaFrance built this 750-gpm pumper on a military chassis. CREDIT: Nelson M. Barter

During World War II several fire apparatus manufacturers used the 1940 International Harvester style chassis. The Bean-Cutler Division of FMC built many 500-gpm pumpers for the military on the International Harvester chassis in San Jose, California. This Marine Corps version appears to be a mid to late 1940s model. CREDIT: Richard Adelman Collection

American LaFrance shipped this 1941 Series B575 Rotary Gear open cab pumper to the Boston Navy Yard, Charlestown, Massachusetts, in November 1941. Serial number L1581 was powered with a Lycoming V-12 190-hp gasoline engine. CREDIT: Richard Adelman Collection

Naval Supply Annex, Stockton, California, used this 1941 American LaFrance 500 Series pumper. CREDIT: Richard Adelman Collection

Norfolk Naval Base firefighters battle a fire in a two-story wood frame structure. In the foreground is a 1940s era Seagrave open cab pumper. CREDIT: Richard Adelman Collection

Long Island Fuel Depot operated this 1941 Type 80 Mack, 500-gpm pumper in Casco Bay, Maine, during World War II. This facility was used to refuel ships during World War II and was disposed of after the war. CREDIT: Nelson M. Barter

This 1942 Seagrave-Ford, powered by a 95-hp V-8, delivered 500 gpm and served the Naval Air Station, Seattle, Washington. Delivered in October 1941, the unit carried 1000 feet of 2 1/2-inch hose, and 150 gallons of water. CREDIT: Bill Killen Collection

U.S. Naval Proving Ground, Dahlgren, Virginia's 1941 Buffalo pumper was featured in Buffalo Fire Appliance Corporation's "Modern Fire Defense" advertisement in fire service trade journals. CREDIT: Peter West

The Naval Air Station where this pair of 1941 Ford Howe 500-gpm pumpers served is unknown. CREDIT: Bill Killen Collection

This Peter Pirsch Model 14 1000-gpm pumper with GMC Cab was delivered to the Norfolk Navy Yard, Portsmouth, Virginia, in April 1941 for a total cost of $10,650.00. CREDIT: Bill Killen Collection

This 1941 Chevrolet with a W.S. Darley pump, water tank, and booster reel was most likely used as a brush truck. CREDIT: Jim Atkinson Collection

This 1941 Ford pickup was used as a Fire Patrol vehicle at the Bethlehem Steel Plant in Hingham, Massachusetts. CREDIT: Jim Atkinson Collection

This 1941 Ford American LaFrance carbon dioxide rig served the Portsmouth Naval Shipyard, Portsmouth, New Hampshire. CREDIT: Jim Atkinson Collection

This Peter Pirsch Model 41 75-foot 2-piece wood Senior Aerial, hoist number H-115, was assigned to the Naval Supply Depot, Norfolk, Virginia. Pirsch chassis serial number 1291 was powered by a Waukesha 145 GK gasoline motor and was delivered July 8, 1942. CREDIT: Bill Killen Collection

This 1941 Seagrave-Ford is believed to have been assigned to the Radio Communications Station, Cheltenham, Maryland. Seagrave records indicate it was assigned to the Naval Air Facility Academy in Annapolis. Quite often the Navy would change the destination of fire apparatus from the factory or reassign apparatus to different Navy bases. CREDIT: Bill Killen Collection

Bethlehem Steel operated the U.S. Naval Shipyard in Quincy, Massachusetts, for the Navy. This 1941 Mack 750-gpm pumper was ready for winter snow conditions as evidenced by the snow chains on the rear tires. CREDIT: Raoul K. Denton Collection

Marine Corps Base Camp Pendleton's pair of 1941 Seagrave pumpers built on International Harvester K5 model chassis. CREDIT: Jim Atkinson Collection

This 1941 Peter Pirsch Model 21 is similar to the Model 21 built for the Navy's Lualualei Ammunition Depot in 1934. This unit was built for the Navy Ammunition Depot, Hawthorne, Nevada, and had serial number 1200. This engine was equipped with a 500-gpm pump. The rims on the 1941 Model 21, although similar in design to the wheels on the 1934 Model 21, did not use pneumatic tires and tubes. CREDIT: Bill Killen Collection

Construction of this 1941 Peter Pirsch Model 20 triple combination began December 12, 1941. The truck carried serial number 1284 and was equipped with a 750-gpm two-stage pump. The truck was delivered to the U.S. Naval Supply Depot, Oakland, California, on March 12, 1942. Bill Killen Collection

Naval Station Pearl Harbor's 1941 Sanford 750-gpm pumper had a 185-gallon water tank and was later sold to Hawaii County, Hawaii. Sanford Fire Equipment Corporation delivered 10 Model N-75 pumpers to Navy installations in California, New Jersey, New York, Oregon, and Rhode Island between 1943 and 1945. CREDIT: Richard Adelman

This 1941 Peter Pirsch Quad saw extensive service with the Navy Yard Fire
Department at Pearl Harbor, Hawaii. CREDIT: Raoul K. Denton Collection

At the beginning of World War II the Navy experimented with modified Indian
motorcycles equipped with a small amount of carbon dioxide and a two-man
crew. CREDIT: Bill Killen Collection

The Portsmouth, New Hampshire, Navy Yard received this Battleship Gray 1942 Seagrave Model 66, 85-foot 4-section aerial ladder in June 1942. CREDIT: Bill Killen Collection

This 1942 Ford American LaFrance 500-gpm pumper served the Puget Sound Navy Yard, Bremerton, Washington. CREDIT: Jim Atkinson Collection

This circa 1942 American LaFrance pumper was assigned to the Naval Ammunition Depot, Earle, New Jersey. CREDIT: Bill Schwartz & Jim Burner Collection

Portsmouth Naval Shipyard Fire Department's pumping units: Engine 2 is an unidentified 500-gpm pumper, possibly a Stewart chassis, Engine 1 is a 1942 1000-gpm Mack, and Engine 3 is a 1942 Seagrave 750-gpm pumper. CREDIT: Portsmouth Naval Shipyard Fire Department

The City of Manchester, New Hampshire, originally purchased this 1942 Mack 1000-gpm pumper. World War II had just started and the U.S. Government needed fire trucks. Upon the orders of President Roosevelt and the War Department, this vehicle and many others were turned over to the government for the war effort. Bought for the sum of $10,376.00 from Mack Truck, Inc., New York, New York, this truck was placed in service at Portsmouth Naval Shipyard, New Hampshire, on March 9, 1942. The civilian fire department had just relieved the Marines. CREDIT: Portsmouth Naval Shipyard Fire Department

This 1942 Seagrave Model 80E 750-gpm pumper carried 200 gallons of water and served the Naval Station in Dutch Harbor, Alaska. CREDIT: Bill Killen Collection

Buffalo Fire Appliance Corporation built this 1942 750-gpm pumper for the U.S. Navy Destroyer Base, San Diego, California. CREDIT: Peter West

Naval Station Pearl Harbor's 1942 Seagrave Model 66-240, 85-foot Quint was identical to the unit delivered to Naval Air Station Seattle, Washington, except the paint on this unit was primer only. CREDIT: Bill Killen Collection

This photograph was taken at the Norfolk Naval Base, Norfolk, Virginia, and shows a 1942 era Peter Pirsch aerial ladder extended with four firefighters at different heights. To the left in the shadows is a 1942 Ford Station Wagon Chief's car. CREDIT: Bill Killen Collection

51

American LaFrance shipped this 1942 Type B-612 carbon dioxide pumper to the Mare Island Naval Shipyard in Vallejo, California, on June 12, 1942. CREDIT: Jim Atkinson Collection

This 1942 American LaFrance Type B-612 carbon dioxide pumper, registration number L-1884, was shipped to the U.S. Naval Ammunition Depot, Iona Island, New York, on September 5, 1942. CREDIT: Bill Killen Collection

American LaFrance built this 6 x 6 firefighting vehicle on a military chassis for the Navy Bureau of Ordnance. CREDIT: Richard Adelman Collection

Civilian firefighters at the U.S. Marine Corps station at Sinaiana, Philippine Islands, operated this 1942 International 500-gpm structural pumper. CREDIT: Raoul K. Denton Collection

This 1942 American LaFrance B-675 open cab pumper L-1638 was delivered to the Washington Navy Yard, Washington, DC, on February 18, 1942. CREDIT: Raoul K. Denton Collection

This 1942 Seagrave 750-gpm pumper carried 150 gallons of water and was assigned to the Naval Ordnance Investigation Laboratory, Stump Neck, Indian Head, Maryland. "Stump Neck" was a remote facility used to train explosives ordnance disposal technicians during World War II. CREDIT: Raoul K. Denton Collection

This 1943 era Ward LaFrance 1250-gpm pumper is painted Haze Gray and was assigned to the U.S. Naval Ammunition Depot, Portsmouth, Virginia. Dick Adelman saw this truck in 1947 after the Navy painted the truck red. CREDIT: Richard Adelman Collection

This 1943 Buffalo was one of two custom pumpers assigned to the Naval Operating Base, Kodiak, Alaska. CREDIT: Raoul K. Denton Collection

This 1943 American LaFrance JOX 100-foot 3-section steel aerial was assigned to the Naval Torpedo Station, Newport, Rhode Island. American LaFrance introduced the cab ahead of the engine aerial ladder trucks in 1940. CREDIT: Raoul K. Denton Collection

This 1943 Seagrave Model 80-500 was special ordered with 4 carbon dioxide cylinders mounted across the frame rail behind the pump. This unit was delivered to a Navy contractor, Consolidated Aircraft, Tucson, Arizona. CREDIT: Bill Killen Collection

This 1943 Model 66 Seagrave served the Naval Hospital, Bethesda, Maryland. The 1000-gpm pumper carried 100 gallons of water. CREDIT: Bill Killen Collection

U.S. Naval Station, Portsmouth, New Hampshire's 1943 Buffalo 500-gpm pumper with a Stewart body. CREDIT: Peter West

U.S. Naval Supply Depot used this 1943 Buffalo Fire Appliance pumper at the Seattle Naval Station, Seattle, Washington. CREDIT: Peter West

This 1943 Buffalo Type 1000 served the U.S. Naval Supply Depot in Bayonne, New Jersey. The doors were removed because the 3-inch valve was in the way. CREDIT: Peter West

This is one of eleven 1943 Buffalo Model 700 pumpers built for the U.S. Navy on International cab and chassis with the distinctive Buffalo grill. CREDIT: Peter West

This is one of many Trailer Fire Pump combinations built for the Navy during World War II by the Peter Pirsch Company. Several manufacturers built trailer mounted pumps for the military up until the 1950s. CREDIT: Bill Killen Collection

Hale Pumps delivered a large number of trailer mounted 500-gpm pumps to the U.S. Navy during World War II. This rear and right side view shows the pump, hard sleeves, axe, and discharge on the side of the trailer. CREDIT: Bill Killen Collection

The Navy purchased more than 60 of these three-piece 65-foot wood ladders for use at the Naval Air Station, Lakehurst, New Jersey. It is believed they were used to support Navy Balloons. One interesting aspect regarding this photograph is that this part of the factory remained virtually unchanged right up to the last period of time that Peter Pirsch remained in business. CREDIT: Roger Bjorge Collection

This American LaFrance engine was assigned to the Fort Mifflin Ammunition Depot, Philadelphia, Pennsylvania. CREDIT: Bill Killen Collection

This 1943 International 6 x 6 served Portsmouth Naval Shipyard, Portsmouth, New Hampshire, as Foam 1. CREDIT: Jim Atkinson Collection

This unidentified Buffalo 500-gpm pumper is circa 1941-1943 and is believed to have been assigned to the Philadelphia Navy Yard, Philadelphia, Pennsylvania. CREDIT: Bill Killen Collection

This 1940 GMC pumper served at an unidentified Marine Corps installation. CREDIT: Jim Atkinson Collection

Naval Station Treasure Island, San Francisco, California, used this 1944 American LaFrance JOX 3-section steel 75-foot midship mounted aerial ladder. CREDIT: J. Chris Scheer

This 1944 American LaFrance JOX aerial ladder served the Philadelphia Navy Yard, Philadelphia, Pennsylvania, until the late 1950s. CREDIT: John J. Wentzel

Buffalo Fire Appliance Corporation's advertisement "Reliable Fire Defense" featured the U.S. Naval Magazine, Indian Island, Washington's pumper as "One of many Buffalo Pumpers protecting U.S. Naval Yards." CREDIT: Peter West

Seagrave built this 1944 pumper on a Studebaker chassis with a 500-gallon tank for the Naval Hospital in Astoria, Oregon. CREDIT: Bill Killen Collection

U.S. Naval Magazine, Port Chicago, California's 1944 Model 66, 65-foot Quint. This open cab unit was equipped with a 750-gpm pump. CREDIT: Bill Killen Collection

The U.S. Naval Mine Depot at Yorktown, Virginia, used this 1944 American LaFrance 610 open cab pumper to protect ordnance storage facilities. CREDIT: Raoul K. Denton Collection

This red 1945 Buffalo two-door open cab quad served the Philadelphia Naval Shipyard, Philadelphia, Pennsylvania, as Ladder 1 until the late 1950s or early 1960s. The Trevose Fire Department in Bucks County, Pennsylvania, purchased the truck from the Navy and repainted the rig yellow. CREDIT: Raoul K. Denton Collection

Buffalo Fire Appliance Corporation's 1945 1000-gpm pumper served the Boston Navy Yard, Charlestown, Massachusetts. CREDIT: Jim Atkinson Collection

Naval Air Test Center Patuxent River, Maryland's 1945 Seagrave 65-foot midship mounted aerial ladder was delivered in November 1945. CREDIT: Bill Killen Collection

These American LaFrance Foamite Corporation hose and ladder carts were used throughout the Naval Powder Factory, Indian Head, Maryland. The hose carts carried approximately 500 feet of 2 1/2-inch hose and an assortment of nozzles and both hydrant and spanner wrenches. The Naval Powder Factory built pump houses on the Potomac River with large volume pumps supplying water for sprinkler systems and extensive fire hydrants located throughout the explosives production facilities. CREDIT: Frank Cotrufo, Jr. Collection

This Yard Tug Boat (YTB) was stationed at the Portsmouth Naval Shipyard, Portsmouth, New Hampshire, and served as their fireboat. CREDIT: Portsmouth Naval Shipyard Fire Department

The U.S. Naval Supply Research and Development Facility in Bayonne, New Jersey, used this 1946 Chevrolet 500-gpm pumper. A 1940s vintage Seagrave open cab pumper can be seen inside the fire station. CREDIT: Raoul K. Denton Collection

This 1946 Seagrave-International 500-gpm pumper was assigned to the U.S. Naval Academy, Annapolis, Maryland. CREDIT: Joe MacDonald

This Dodge Power Wagon style military brush firefighting truck was assigned to the U.S. Naval Academy and is believed to be a 1941 or 1942 model. CREDIT: Ed Bosanko Collection

This 1949 Ford pickup served as the Fire Chief's vehicle at the Naval Powder Factory, Indian Head, Maryland. The pumper in the background is a 1940 GMC. CREDIT: Frank Cotrufo, Jr. Collection

This John Bean High Pressure Fog on an International Harvester 4 x 4 delivered 60-gpm, 2,800 pounds per square inch. The Navy base where this unit was located is unknown. CREDIT: Bill Killen Collection

Floyd Bennett Naval Air Station in Brooklyn, New York, used this 1949 Reo tanker for water supply. CREDIT: Fire Apparatus Journal Collection

The manufacturer of this 750-gpm pumper on a 1958 or 1959 International Harvester chassis is unknown. This vehicle was assigned to St. Albans Naval Hospital in New York. CREDIT: Charlie Bowman

The Marine Corps Base where this 1949 Ward LaFrance was assigned is unknown. CREDIT: Richard Adelman Collection

U.S. Naval Academy Fire Marshal and fire inspectors pose with their 1950 Chevrolet pickups and fire prevention float. CREDIT: Doug Howard

This John Bean High Pressure Fog trailer could be pulled by a pickup or by hand. The trailer was equipped with a pump mounted behind the front wheel, a booster reel above the pump, and a water tank (water capacity unknown). CREDIT: U.S. Navy

The Navy purchased several 750-gpm fire engines from the Four Wheel Drive Auto Company, Clintonville, Wisconsin, in 1950-1951. CREDIT: Bill Killen Collection

General Detroit built many types of vehicles for the military, including 750-gpm pumpers on Federal chassis in the early 1950s. This 1954 General Detroit-Federal 750-gpm pumper was assigned to the U.S. Naval Construction Battalion, Gulfport, Mississippi. CREDIT: Raoul K. Denton Collection

This 1952 American LaFrance 700 Series 65-foot junior aerial was assigned to Naval Air Station Quonset Point, Rhode Island. CREDIT: Richard T. DeUso

This 1952 American LaFrance 700 Series junior 75-foot 3-section steel aerial ladder was shipped to Naval Submarine Base New London, Groton, Connecticut, on November 28, 1952. The open cab model was painted red and powered with a Continental gasoline engine and a manual transmission. The truck carried 228 feet of ground ladders. CREDIT: Bill Killen Collection

This 1952 Autocar tractor and 3000-gallon water tanker provided a mobile water supply to remote areas of the Naval Ammunition Depot, Earle, New Jersey. CREDIT: Bill Schwartz & Jim Burner Collection

Brunswick Naval Air Station, Brunswick, Maine, fire department personnel rotated duty assignments at the Harpswell Fuel Depot where this 1952 GMC Foam Truck was assigned. CREDIT: Nelson M. Barter

The Harpswell Fuel Depot supplied aviation fuel to Brunswick Naval Air Station via pipeline. Fuel was transferred from ships to eight 80,000-gallon and six 50,000-gallon above ground storage tanks. This 500-gpm pumper, built on a 1955 International R-180 chassis, was assigned to Harpswell Fuel Depot until about 1968 when the Navy ceased using AVGAS for the P2V Neptune patrol aircraft. CREDIT: Nelson M. Barter

This picture was taken at Floyd Bennett Naval Air Station in Brooklyn, New York. It is believed that the Seagrave quad was originally assigned to the Brooklyn Navy Yard, Brooklyn, New York. CREDIT: Fire Apparatus Journal Collection

This fleet of fourteen 1953-1954 FWD pumpers were photographed at the FWD plant in Wisconsin, March 1954 prior to delivery to the U.S. Navy. CREDIT: Bill Killen Collection

Riding the tailboard of this 1953 General Detroit-Federal 750-gpm pumper during the Naval Ordnance Station Fire Department's 1960 Fire Prevention Week Parade is the author. CREDIT: Carole Killen

This 1953 Ward LaFrance Model CW 750-gpm pumper carried 175-gallons of water and 40-gallons of foam and was assigned to the U.S. Naval Radio Transmitting Facility, Annapolis, Maryland. CREDIT: J. MacDonald

Raytheon Corporation was one of many contractors that operated government production plants for the Navy. This 1953-1954 era Ward LaFrance 750-gpm pumper was assigned to an unidentified Raytheon plant. CREDIT: Bill Killen Collection

This 1954 American LaFrance 700 Series pumper was assigned to the Boston Navy Yard, Charlestown, Massachusetts. This pumper remained in service for more than twenty years and was manned for more than 72 consecutive hours at the Chelsea, Massachusetts, conflagration in 1973. CREDIT: Bill Killen Collection

The U.S. Navy experimented with a four-wheel drive "advanced base fire pumper" in 1954. This prototype was believed to have been built by Ward LaFrance. This 750-gpm pumper carried 200 gallons of water, 30 gallons of foam, and 30 gallons of wet water. CREDIT: Bill Killen Collection

FWD built several 1954 brush tankers with 240-gpm pumps and 400-gallon water tanks. The only known survivor from this series of FWD brush trucks is at the former Naval Academy Dairy farm in Millersville, Maryland. CREDIT: Bill Killen Collection

The U.S. Marine Corps ordered several 1954 era Ward LaFrance 750-gpm pumpers like this one. CREDIT: Richard Adelman Collection

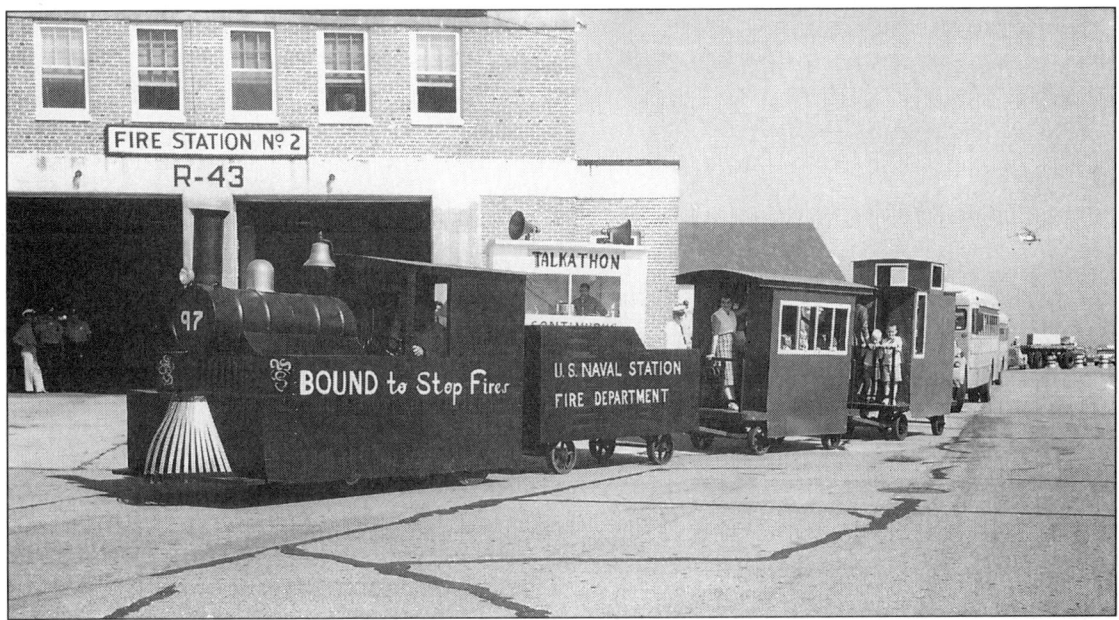

Norfolk Naval Station firefighters obtained farm tractor and bomb carts from surplus and constructed this Fire Prevention Train in the early 1950s. CREDIT: U.S. Navy

American LaFrance built the U.S. Navy's YDXF-1 foam pumper. This experimental one-of-a-kind was powered with a 6-cylinder 820-cubic-inch 320-hp gasoline engine and a heavy-duty 5-ton military manual transmission. CREDIT: Ed Peterson Collection

Based on the USN number on the door, Oceana Naval Air Station, Virginia Beach, Virginia's Ward LaFrance is probably a 1954 or 1955 era 750-gpm pumper. CREDIT: U.S. Navy

Naval Air Station Johnsville (Warminster) fire department conducts hose evolutions with their 1955 Ward LaFrance 500-gpm pumper. A 1961 Dodge pickup served as the Fire Chief's vehicle. CREDIT: Raoul K. Denton Collection

This 1956 800 Series American LaFrance 750-gpm pumper served the U.S. Naval Station, Treasure Island, San Francisco, California. CREDIT: Bill Killen Collection

The location and number of these 1956 American LaFrance Model 7-85 ALO aerial ladders delivered to the Navy is unknown. The Navy purchased American LaFrance ladders in 1952, 1965, and 1974. CREDIT: Bill Killen Collection

This 1956 Mack open cab 750-gpm pumper was assigned to the Naval Air Propulsion Center, Trenton, New Jersey. CREDIT: Raoul K. Denton Collection

This 1956 International Harvester Brush truck was assigned to the Naval Powder Factory, Indian Head, Maryland. CREDIT: U.S. Navy

Naval Station Crane, Indiana, firefighters mop-up after a flue fire in Navy housing in the mid-1950s. Their 1953 Detroit General-Federal 750-gpm pumper can be seen in the foreground. CREDIT: U.S. Navy

This 1957 International-General 750-gpm pumper with a 500-gallon tank was assigned to the Naval facility at Vieques, Puerto Rico. CREDIT: John Richardson

Naval Air Station Oceana fire department's 1957 International open cab pumper. This 750-gpm midship pump carried 200-gallons of water. This truck was one of several delivered to the Navy for $10,275.00 each. CREDIT: U.S. Navy

The on-duty crew poses with their 1957 GMC pumper. Fire Station 15 was at Naval Ammunition Depot Lualualei, on Oahu, Territory of Hawaii. CREDIT: U.S. Navy

Fire Chief Hank Vescovy is shown communicating via his bullhorn as more than 150 fire engines operate at draft from piers at Submarine Base New London, Groton, Connecticut. CREDIT: James Manser, Retired Fire Marshal Administrator

YTB 548 and 150 pumpers move tons of water at Naval Submarine Base, Groton, Connecticut, in 1957. CREDIT: James Manser, Retired Fire Marshal Adminstrator

Naval Air Development Center, Warminster, Pennsylvania, used this 1957 Ford 2000-gallon water tanker in support of airfield operations. CREDIT: Raoul K. Denton Collection

The fire department at Brunswick Naval Air Station, Brunswick, Maine's water distributor/tanker is a late 1940s or early 1950s vintage Autocar and is shown here in response to a passive defense drill on March 22, 1957. CREDIT: U.S. Navy

Norfolk Naval Station fire department's 1958 B Model Mack operating at a fire at the Naval Air Rework Facility Engine Test Cell on February 11, 1976. CREDIT: Danny Miller Collection

Sailors conduct training evolutions with the YTB fireboat at the Naval Ship Research and Development Center, Annapolis, Maryland, circa 1950s. CREDIT: Ed Bosanko Collection

This 1958 Central International Harvester Brush Structural pumper was transferred to the Marine Corps Base Quantico from the Naval Station, Dahlgren, Virginia. This yellow rig had a 500-gpm pump and carried 600 gallons of water. CREDIT: Jim Atkinson Collection

This American LaFrance Series 700 open cab 3-section midship mounted aerial ladder was assigned to the Canal Zone. CREDIT: Chris Scheer

Bob Jennings took this photograph of the U.S. Naval Academy's 1959 Detroit General-Federal in front of the Naval Academy's old fire station, Annapolis, Maryland. CREDIT: Jim Atkinson Collection

This 1959 International General 500-gpm Brush Structural rig carried 500 gallons of water and was assigned to the Naval Facility at Vieques, Puerto Rico. CREDIT: John Richardson

This 1959 Cardox 0-6 crash truck is built on a 2 1/2-ton chassis with 4000 pounds of carbon dioxide. A 203-hp Continental six-cylinder gasoline engine powered this rig. The Navy base where this unit was assigned is unknown. CREDIT: Bill Killen Collection

Naval Air Station Johnsville fire department's 1959 International Dakota pumper. Naval Air Station Johnsville, Warminster, Pennsylvania, was renamed the Naval Air Development Center. CREDIT: Raoul K. Denton Collection

The venerable Jeep provided a fast response firefighting dry chemical unit at Marine Corps Air Stations. This 1952-1953 era vintage jeep carried 300 pounds of Ansul Dry Chemical. CREDIT: Richard Adelman Collection

Fire Trucks Inc. built several of these 500-gpm pumpers on a 1960 GMC chassis for the U.S. Navy. The Navy base where this unit was assigned is unknown. CREDIT: Jim Atkinson Collection

The Washington Navy Yard fire department, Washington, DC, used this early 1960s GMC 500-gpm structural pumper. CREDIT: Raoul K. Denton Collection

The fire department at the Naval Training Center Bainbridge, Maryland, used this 1960 International-Seagrave 750-gpm pumper. CREDIT: Bob Kimbal

Walter Kidde manufactured this airfield firefighting vehicle on a GMC 6 x 4 chassis for the Navy. The Gordon's Corners Volunteer Fire Department purchased this truck in the early 1960s. CREDIT: John Rieth Collection

FWD Corporation's Truck, Airplane Firefighting and Rescue, Foam and Water 6 x 6 Type MB-1A was demonstrated (although never produced) for the Navy on August 29, 1961 at the FWD Proving Ground in Clintonville, Wisconsin. The traction engine was a Continental six-cylinder gasoline engine and had a top speed (governed) of 61 mph and accelerated from 0-50 mph in 45 seconds. The pumping engine was Chrysler HT31 gasoline eight-cylinder "V" with a displacement of 361 cubic inches. The main foam pump was manufactured by Cardox and had a discharge rate of 6000 gpm. CREDIT: Bill Killen Collection

The Navy purchased several 1962 Brush-Structural pumpers on International Harvester chassis' at a cost of $12,432.00. These trucks were equipped with 500-gpm pumps and carried 600 gallons of water. CREDIT: U.S. Navy

This 1962 FWD P2 crash truck was remanufactured in 1981 by Quality Manufacturing in Alabama and assigned to Naval Air Station North Island, Coronado, California. CREDIT: Bill Killen Collection

This early 1960s Model 530B Brush Structural pumper was built by Ward LaFrance on a Reo chassis. The 500-gpm pump was supplied from a 500-gallon water tank. CREDIT: Jim Atkinson Collection

The installation where this 1963 Dodge twinned agent unit is assigned is not known. As a result of the research by Dr. Richard Tuve, Naval Research Laboratory, Washington, DC, the twinned agent unit played a significant role in airfield fire protection, particularly at smaller airfields and as a quick attack unit. CREDIT: Jim Atkinson Collection

This 1963 Fire Trucks Inc.-International was assigned to the U.S. Naval Amphibious Base, Little Creek, Virginia. This yellow rig was equipped with a 750-gpm pump and carried 350 gallons of water and 30 gallons of Aqueous Film Forming Foam. CREDIT: Raoul K. Denton Collection

The Naval Underwater Systems Command fire department in New London, Connecticut, used this 1963 Fire Trucks Inc. 750-gpm pumper. CREDIT: John Hannan, Navy Fire Marshal, Retired

Naval Surface Warfare Center Dahlgren, Virginia's 1969 Oshkosh MB-5 Crash Truck. CREDIT: U.S. Navy

Norfolk Naval Base fire department's 1964 General Safety 500-gpm Brush Structural pumper carried 400 gallons of water. Taken out of front line service in the early 1980s, this truck was used for testing fire hose for several years. CREDIT: Jim Atkinson Collection

Naval Air Station firefighters in Lakehurst, New Jersey, built this "Brush Breaker" on a 1964 White-Reo 6 x 6 military chassis. The firefighting package is a 250-gpm portable pump and a 600-gallon water tank. CREDIT: Jim Atkinson Collection

This 1964 Fire Trucks Inc.-International 750-gpm pumper was assigned to the Portsmouth Navy Shipyard, Portsmouth, New Hampshire. In 1977 the rig was transferred to the Naval Air Propulsion Center fire department in Trenton, New Jersey. CREDIT: Raoul K. Denton Collection

This 1964 American LaFrance 900 Series 100-foot aerial tractor trailer combination served the Philadelphia Naval Shipyard, Philadelphia, Pennsylvania, until 1980 when it was replaced with a 1980 Spartan Pierreville 100-foot rear mount. CREDIT: Raoul K. Denton Collection

This 1965 GMC 750-gpm pumper was assigned to Naval Training Center Bainbridge, Maryland. CREDIT: Bob Kimball

Naval Station Pearl Harbor fire department's 1965 American LaFrance Pioneer escorts one of the Apollo space capsules from the pier at the Pearl Harbor Naval Station, Honolulu, Hawaii, to a receiving facility for shipment to the Kennedy Space Center. CREDIT: Victor Flint

This 1965 American LaFrance 100-foot tiller was assigned to Naval Training Center Great Lakes, Chicago, Illinois. CREDIT: Ron Heal

Naval Ordnance Station firefighters man handlines from the Indian Head Volunteer Fire Department's 1965 American LaFrance Pioneer pumper at a fire in the Navy Dispensary, Indian Head, Maryland. CREDIT: Frank Cotrufo Jr.

This yellow Fire Trucks Inc. 750-gpm pumper was assigned to the Cheatham Annex in Yorktown, Virginia. Built on an International chassis, the truck carried 300 gallons of water. CREDIT: Jim Atkinson Collection

This red 1965 Progress Industries-International Harvester 500-gpm Brush Structural rig served the Naval Research Laboratory, Chesapeake Beach, Maryland. The Naval Research Laboratory Fire Department is the only career fire department in Calvert County, Maryland, and provides mutual aid to several volunteer fire companies. CREDIT: Bill Killen Collection

Keflavik Naval Air Station, Keflavik, Iceland, received this 1974 American LaFrance 85-foot, 3-section steel midship mounted aerial ladder from Naval Station Norfolk in 1993. This rig was one of three Navy units delivered to the Navy in 1974, each equipped with a Detroit Diesel 6-71 engine and 5-speed manual transmission. Each truck carried a complement of 208 feet of ground ladders. CREDIT: Bill Killen

Naval Air Station South Weymouth, Massachusetts' 1965 International crash truck carried 4500 gallons of water, 500 gallons of foam, and was equipped with a 500-gpm pump. CREDIT: Mark A. Redman

This 1966 Progress Industries International Harvester served the Submarine Base New London, Groton, Connecticut. The 500-gpm Brush Structural pumper carried 600 gallons of water and was painted red. CREDIT: Jim Atkinson Collection

Naval Communication Station Cheltenham, Maryland, used this 1966 Fire Trucks Incorporated (Fire Trucks Inc.) 750-gpm pumper until 1986. Fire Trucks Inc. built several units using this International R-185 chassis in 1964-1968. The units were equipped with an International Harvester gasoline engine, manual transmission, Waterous 750-gpm pumps, and a 300-gallon water tank. CREDIT: Bill Killen Collection

Naval Base Norfolk's 1966 American LaFrance Pioneer operating at a fire at the Naval Rework Facility Test Cell fire on February 11, 1976. CREDIT: Danny Miller Collection

Naval Air Station Whiting Field in Milton, Florida, is the Navy's basic training school for fixed wing and helicopter pilots. In addition to the main runways at Whiting Field, the Navy operates several outlying auxiliary fields in northern Florida and southern Alabama. This 1966 International 5000-gallon tanker was used to re-supply crash trucks at Whiting Field. CREDIT: Raoul K. Denton

Norfolk Naval Shipyard, Portsmouth, Virginia, firefighters seen here removing a ladder from a 1966 Fire Trucks Inc. 750-gpm pumper at a building fire. CREDIT: U.S. Navy

Norfolk Naval Shipyard's two 1966 Fire Trucks Inc. 750-gpm pumpers operating at a building fire at Norfolk Naval Shipyard, Portsmouth, Virginia. CREDIT: U.S. Navy

Built on a 1967 International Harvester chassis, this lime green 1967 Progress Industries 500-gpm Brush structural rig served the Naval Surface Warfare Center, White Oak, Silver Spring, Maryland. CREDIT: Bill Killen Collection

This 1967 Fire Trucks Inc.-International 750-gpm pumper served as a reserve engine to several Navy installations in the San Diego, California, region. CREDIT: Bill Killen Collection

Norfolk Naval Base, Norfolk, Virginia, used this 1967 Kaiser-MacLeod 6000-gallon tanker for runway foaming and support for airfield operations. CREDIT: Jim Atkinson Collection

This 1967 Ward LaFrance 65-foot midship 2-section aerial ladder was assigned to Naval Air Station, North Island, Coronado, California. CREDIT: Bill Killen Collection

This 1967 Kaiser 6 x 6 American Air Filter was assigned to the U.S. Marine Corps airfield at Quantico, Virginia. The Model 530C had a 750-gpm pump, and carried 400 gallons of water and 40 gallons of Aqueous Film Forming Foam. CREDIT: Jim Atkinson Collection

This 1967 Kaiser 6 x 6 brush truck is equipped with a 200-gpm pump and an 800-gallon water tank and is assigned to the Camp Pendleton Marine Corps Base, Camp Pendleton, California, fire department. CREDIT: Jim Atkinson Collection

Firefighters employed at the Naval Communications Station Cheltenham, Maryland, built this brush rig on a 1967 Willys Jeep. The firefighting package is a 250-gpm portable pump and a 200-gallon water tank. Credit: Bill Killen Collection

Chief Jim Derstine photographed Long Beach Naval Station, Long Beach, California, fire department's 1968 American LaFrance Pioneer 1000-gpm pumper. CREDIT: Jim Derstine

This 1968 Macleod-GMC tanker is still in service at the Naval Weapons Station Yorktown, Virginia. CREDIT: U.S. Navy

U.S. Naval Construction Battalion Gulfport, Mississippi, used this 1968 General Safety Equipment 750-gpm pumper on an International Harvester R185 chassis. CREDIT: Raoul K. Denton Collection

Fire department personnel assigned to Keflavik Naval Air Station, Keflavik, Iceland, modified this 1968 Fire Trucks Inc.-International Harvester Brush Structural pumper to support airfield operations. CREDIT: Bill Killen

Marine Corps Air Station Beaufort (South Carolina) fire department's 1968 Fire Trucks International GMC 750-gpm pumper. The Marine Corps repainted all of their fire apparatus fleet to red with a white stripe in the mid-1980s. CREDIT: Jim Atkinson Collection

Naval Air Station Bermuda fire department operated this 1968 GMC-Macleod 2000-gallon tanker in support of airfield operations. CREDIT: Raoul Denton

Oceana Naval Air Station, Virginia Beach, Virginia, fire department's 1968 Fire Trucks Inc. 750-gpm pumper was painted red. The Navy instituted a new paint scheme for fire apparatus in 1985 when this truck was repainted lime green with white cab roof and white reflective stripe. CREDIT: Jim Atkinson Collection

Naval Air Station Patuxent River, Lexington Park, Maryland, used this 1968 Progress Industries-International Harvester Brush Structural pumper at Webster Field. The 500-gpm pumper carried 600 gallons of water. CREDIT: Jim Atkinson Collection

This 1969 Fire Trucks Inc.-GMC 750-gpm pumper served Naval Amphibious Base Coronado, California. This photograph was taken in the early 1980s after the consolidation of San Diego area Navy fire departments into the Federal Fire Department San Diego. CREDIT: Garry Kadzielawski

San Diego Naval Shipyard, San Diego, California's Engine 7 is a 1969 Fire Trucks Inc.-GMC 750-gpm pumper. The self-contained breathing apparatus are mounted above the compartments on the left side of this lime green truck. The water tank carries 300 gallons of water. CREDIT: Garry Kadzielawski

This 1969 Fire Trucks Inc.-GMC 750-gpm pumper served Naval Air Station Whiting Field, Milton, Florida. This truck was replaced with a 1984 Walters-Duplex 1000-gpm pumper in early 1985. CREDIT: Jim Atkinson Collection

The Fleet Combat Training Center at Dam Neck, Virginia, used this 1969 Fire Trucks Inc.-GMC 750-gpm pumper. CREDIT: Jim Atkinson Collection

This 1969 Fire Trucks Inc.-GMC is Naval Air Station Pensacola's reserve engine. The white over lime green paint scheme was added in 1986. CREDIT: Jim Atkinson Collection

Naval Air Station Glenview (Illinois) fire department's Engine 11 is a 1969 Fire Trucks Inc.-GMC. This red 750-gpm pumper has a 300-gallon tank, a booster reel mounted above the pump compartment and self-contained breathing apparatus and fire extinguishers mounted above the left side compartments. CREDIT: Jim Atkinson Collection

The Naval Ordnance Station, Indian Head, Maryland, used this 1969 Fire Trucks Inc.-GMC to protect explosive manufacturing facilities. This truck equipped with a 750-gpm pump and a 450-gallon water tank was later assigned to Station 2 at Stump Neck Annex. CREDIT: Jim Atkinson Collection

Naval Weapons Station Earle's circa 1970 Ford 3000-gallon tanker. This unit had a 500-gpm pump and Stang gun mounted above the pump. CREDIT: John Rieth

Naval Air Station Oceana fire department's 1978 Dodge Fire Tec rescue vehicle. CREDIT: Bill Killen Collection

Firefighter Marvin Peterson is all smiles, as children from the local elementary school visit the Naval Ordnance Station fire department in Indian Head, Maryland, for a tour of the fire station and a climb on the fire engine. CREDIT: Frank Cotrufo, Jr.

Gibson manufactured seven of these 5000-gallon runway foam trailers for the Navy at a cost of $29,075.00 each in 1975. The installation where this foam trailer was initially assigned is unknown, however, from February 1986 until April 1993 it was assigned to Naval Air Station Sigonella, Italy. CREDIT: Bill Killen Collection

Miramar Naval Air Station, San Diego, California, firefighters demonstrate the pump and roll capability of their 1964 Brush Structural pumper. CREDIT: U.S. Navy

This 1971 Fire Trucks Inc.-International Harvester carried 1000 gallons of 6 percent foam and was equipped with a 1000-gpm pump. This unit was assigned to Naval Education & Training Center Newport to protect the Fleet Industrial Supply Center Fuel Depot in Newport, Rhode Island. CREDIT: Jim Atkinson Collection

This 1971 Fire Trucks Inc.-GMC 750-gpm pumper served the U.S. Marine Corps Base Twentynine Palms, California. Portable fire extinguishers and self-contained breathing apparatus were mounted above the compartments on the left side of the truck. The rig carried 300 gallons of water. CREDIT: Raoul K. Denton Collection

This 1971 Fire Trucks Inc. Brush Structural rig was built on an International Harvester chassis. The red 500-gpm rig was assigned to Boca Chica Key (part of the Key West Naval Air Station) and carried 600 gallons of water. CREDIT: Garry Kadzielawski

Naval Security Group Northwest, Chesapeake, Virginia, used this 1971 Progress Industries International Harvester Brush Structural pumper. The truck was equipped with a 500-gpm pump and carried 500 gallons of water. CREDIT: Jim Atkinson Collection

This is Marine Corps Base Camp LeJeune, North Carolina's Engine 2. This 750-gpm pumper carried 300 gallons of water with a monitor nozzle mounted above the pump compartment. Portable fire extinguishers and self-contained breathing apparatus were mounted on the hose bed above the left side compartments. CREDIT: Raoul K. Denton Collection

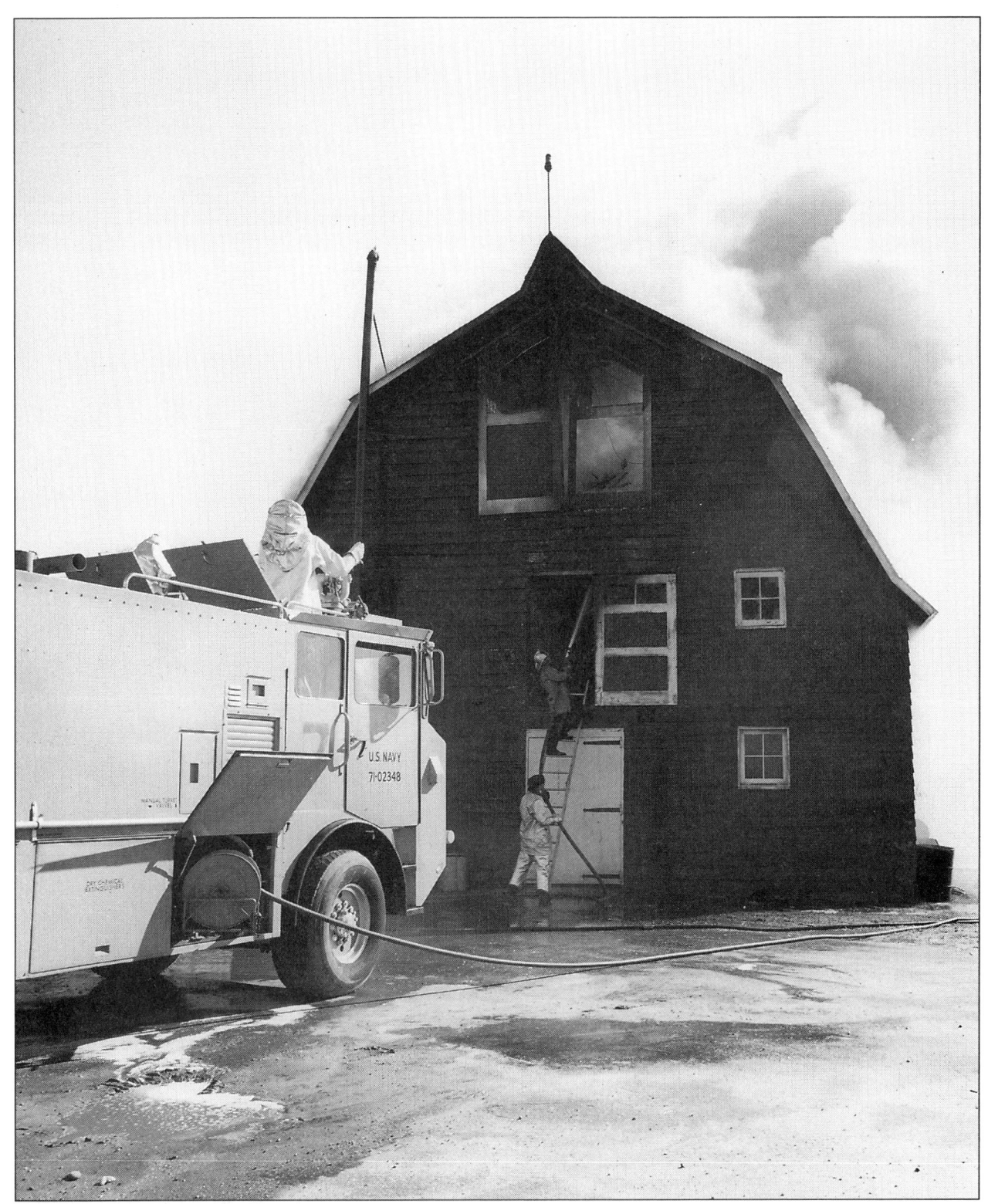

Naval Air Station Patuxent River, Lexington Park, Maryland, assists Saint Mary's County volunteer fire departments at a barn fire in the 1970s. Navy firefighters man two 250-gpm turrets on a 1971 Oshkosh MB-1 as firefighters overhaul the fire. The Navy purchased several MB-1s from Oshkosh Truck for $57,165.00. This MB-1 was taken out of service in 1986 when it was replaced with a 1985 Oshkosh P-19. CREDIT: U.S. Navy

Naval Air Station Key West's Engine 8 is a 1972 Fire Trucks Inc. 50-gpm pumper built on a two-door cab GMC chassis. This unit was overhauled and repainted white over lime green by Quality Manufacturing in 1986. The water tank capacity was 300 gallons. CREDIT: Glenn Vincent

Marine Corps Base Camp LeJeune's 1973 Ward LaFrance-Chevrolet was equipped with a 750-gpm pump and carried 300 gallons of water. CREDIT: Raoul K. Denton Collection

TRUCK, FIREFIGHTING, BRUSH/GRASS

Fire Trucks Inc. built this 1973 Brush Structural pump for the Naval Air Station Bermuda at a cost of $18,733.00. Built on an International Harvester chassis and painted red, this truck had a 500-gpm pump and carried 600 gallons of water. CREDIT: Bill Killen Collection

This 1973 Fire Trucks Inc.-GMC 750-gpm pumper was modified for use as a Squad at Marine Corps Base Camp LeJeune, North Carolina. CREDIT: Raoul K. Denton Collection

This 1973 Ward LaFrance 750-gpm pumper was built on a two-door cab Chevrolet chassis and served the Naval Training Center San Diego, California, as Engine 15. This red pumper carried 300 gallons of water and self-contained breathing apparatus mounted above the compartments on the left side of the rig. CREDIT: Garry Kadzielawski

This 1973 American Motors 530C crash truck was assigned to the Marine Corps at Naval Air Station Millington, Tennessee. This yellow all-wheel drive tandem axle rig was equipped with a 750-gpm pump, a manually operated roof turret, 400-gallon water tank and a 40-gallon foam tank. CREDIT: Raoul K. Denton Collection

The Marine Corps installed a slide-in twin agent firefighting package on this 1973 Chevrolet pickup. This rapid response vehicle carried 400 pounds of Purple-K-Powder and 50 gallons of light water. CREDIT Jim Atkinson Collection

This 1974 Oshkosh P-4A served Naval Air Station Patuxent River, and the Naval Air Test School in Lexington Park, Maryland. CREDIT: Bill Killen Collection

Naval Air Station Patuxent River's 1974 Walters 3000-gallon crash truck served Washington National Airport in Washington, DC, until it was transferred to the Navy sometime around 1986 or 1987. Originally painted red, the Navy repainted the truck lime green during a maintenance overhaul. CREDIT: Bill Killen Collection

The U.S. Naval Academy's 1974 Fire Trucks Inc. custom 750-gpm pumper carried 500-gallons of water and 20-gallons of foam. This unit served several Navy fire departments as a Reserve Engine until it was disposed of in 1997. CREDIT: Bill Killen Collection

This 1974 Flextrac Nodwell firefighting vehicle was assigned to Williams Field, McMurdo, Antarctica. The firefighting package was Halon 1211 and dry chemical. CREDIT: Jim Atkinson Collection

National Naval Medical Center Bethesda's Fire Prevention Bureau used several of these 1975 Cushman mini vehicles to perform fire extinguisher maintenance and building inspections. CREDIT: Bill Killen Collection

The Naval Research Laboratory's Chesapeake Beach Detachment's 1975 Fire Trucks Inc. 750-gpm pumper carried 500 gallons of water. This lime yellow rig was used to provide standby support during fire research activities. This single engine company Navy fire department is the only paid fire department in Calvert County, Maryland, and provides substantial mutual aid response to the county. CREDIT: Bill Killen Collection

Engine 12, Naval Ocean Systems Command San Diego is a 1975 Fire Trucks Inc.-PEMFAB 750-gpm pumper with a 500-gallon water tank and a 20-gallon foam tank. CREDIT: Garry Kadzielawski

Naval Air Station Lakehurst's 1975 Fire Trucks Inc. 750-gpm pumper was modified by firefighter personnel to give it a "Mack" look by installing sheet metal on the front of the cab. CREDIT Bill Killen Collection

This 1975 Kaiser/Gibson runway foamer carried 5000 gallons of water and 800 gallons of foam. This unit was assigned to Naval Air Station Bermuda and was replaced with an Oshkosh P-15 crash truck in 1983. CREDIT: Raoul K. Denton Collection

This red 1975 Maxim 85-foot ladder, on a Mack chassis, served Marine Corps Base Quantico, Virginia, as Ladder 1. This ladder was repainted white over red in the mid-1980s. CREDIT: Bill Killen Collection

This 1975 Dodge with Reading utility body was designated Rescue 6 and assigned to the San Diego Naval Shipyard fire department, San Diego, California. This rescue unit later served multiple stations in the consolidated Federal Fire Department, San Diego. CREDIT: Garry Kadzielawski

This fire in the Officers Club at the Norfolk Naval Shipyard occurred sometime in the early 1970s. As a result of fire damage, the first floor was removed when the building was repaired. CREDIT: Bill Killen Collection

Philadelphia Navy Shipyard's Engine 5 was a 1975 Spartan-Fire Trucks Inc. 750-gpm pumper and carried 500 gallons of water. CREDIT: Jim Atkinson Collection

Gibson built this 1976 foam re-supply unit for the Navy. The firefighting package consisted of a 500-gpm pump, a foam nozzle, and a runway foaming boom mounted on the rear bumper. CREDIT: Bill Killen Collection

Naval Ocean Systems Command's (San Diego, California) Rescue 12. This yellow 1976 Dodge Rescue unit served several Navy installations in the Federal Fire Department, San Diego. CREDIT: Garry Kadzielawski

This 1976 Fire Trucks Inc.-PEMFAB served the Naval Regional Medical Center, San Diego, California. This lime yellow rig had a 750-gpm pump and carried 400 gallons of water. CREDIT: Garry Kadzielawski

Norfolk Naval Shipyard firefighters use the monitor off their 1976 Fire Trucks Inc. 750-gpm pumper during live fire training exercises. CREDIT: U.S. Navy

Naval Surface Warfare Center White Oak, in Silver Spring, Maryland, developed this remote control tracked vehicle for use in firefighting operations involving explosives and munitions manufacturing facilities. The concept was to connect 2 1/2-inch hose lines to the vehicle and advance the vehicle dragging the hose and discharge agent through the nozzle. It worked well in demonstrations but was never perfected and placed into the shore installation fire programs. CREDIT: Bill Killen Collection

This 1977 Seagrave 750-gpm pumper served the Naval Air Development Center fire department in Warminster, Pennsylvania. CREDIT: Raoul K. Denton Collection

This 1977 Fire Trucks Inc.-International Harvester POL (Petroleum, oil, liquid) is one of three units built for the Navy at a cost of $46,292.00. CREDIT: Bill Killen Collection

This 1978 Seagrave Rear Admiral 100-foot 4-section steel rear mounted aerial ladder was one of two units delivered to the Navy in 1978. The drive train was a Detroit Diesel 6-71 265-hp diesel with an Allison Automatic HT-740 transmission. The ladder carried a complement of 193 feet of ground ladders. This truck was remanufactured in 1997 and transferred to the Naval Station Newport, Rhode Island. CREDIT: Jim Atkinson Collection

"Heavy Metal" is the handle Naval Weapons Station Earle firefighters put on this 1977 Dodge 880 4 x 4-brush truck. The firefighting package consists of a 200-gpm pump and a 250-gallon water tank. CREDIT: John Rieth

Bethesda Naval Hospital fire department in Bethesda, Maryland, used this 1977 Dodge four-door crew cab twinned agent unit for helicopter standby support and vehicle fires. CREDIT: Jim Atkinson Collection

Four of the original Oshkosh P4A crash trucks delivered to the Navy in 1976-1977 remain in service in the Navy fire apparatus fleet. Federal Fire Department, San Diego's 1977 Oshkosh P4A was remanufactured by Crash Rescue Equipment Company and is assigned to North Island Naval Air Station, Coronado, California. CREDIT: Jaimie Wood

This 1978 Seagrave 750-gpm pumper served Naval Air Station Patuxent River, Lexington Park, Maryland. The yellow painted rig carried 500 gallons of water and 60 gallons of foam. CREDIT: Bill Killen Collection

Fire Trucks Inc. built this 1978 Brush Structural pumper for the Naval Surface Weapons Center White Oak Laboratory, Silver Spring, Maryland. Painted yellow, this combination truck had a 250-gpm pump and carried 500 gallons of water. CREDIT: U.S. Navy

This 1978 Dodge Fire Tec served as Rescue 12 at the Naval Air Station Key West, Florida, for several years. CREDIT: Glenn Vincent

Naval Shipyard, Pearl Harbor's 1978 Seagrave in action at a building fire sometime in 1979 in Honolulu, Hawaii. CREDIT: Victor Flint

Naval Education and Training Center Newport, Rhode Island's 1978 Seagrave 100-foot aerial was replaced in August 1997 with a rehabbed 1978 Seagrave from Pensacola Naval Air Station, Pensacola, Florida. CREDIT: Bill Killen Collection

This 1978 Seagrave 1000-gpm pumper was one of two ordered with 35-foot Tele-Squrt booms. This unit was assigned to the Naval Ordnance Station, Indian Head, Maryland, and the other unit was assigned to Naval Air Station Bermuda. CREDIT: Bill Killen Collection

This is Marine Corps Base Camp LeJeune's 1978 Seagrave 1000-gpm pumper. This white over red pumper carried 750 gallons of water and 100 gallons of foam. CREDIT: Jim Atkinson

This 1979 Pierce Dodge and Firefly II gas turbine fire pump was assigned to Naval Air Station North Island, Coronado, California. Powered by a gas turbine helicopter engine, the Firefly II delivered 2500 gpm. CREDIT: Garry Kadzielawski

Norfolk Naval Shipyard's 1979 Ward LaFrance P-80 closed canopy cab was one of four units delivered to the Navy. Painted yellow, these Ambassador models featured Maxim 100-foot 4-section steel ladders, tandem axles, and 208 feet of ground ladders. CREDIT: U.S. Navy

This 1979 Ward LaFrance/Maxim 100-foot aerial ladder from the former Subase, New London, was repainted and lettered for the Naval Ship Systems Engineering Station, Philadelphia, Pennsylvania. CREDIT: John M. Calderone

The location where this 1979 Fire Trucks Inc.-GMC Brush Structural pumper was assigned is unknown. This yellow rig is equipped with a 250-gpm pump and a 500-gallon water tank. CREDIT: John Rieth

This 1979 Seagrave 1000-gpm pumper is identical to the units purchased by the Navy and is one of 21 units ordered by the Marine Corps. Originally painted yellow, the Marine Corps repainted their fire apparatus fleet to white over red in the mid-1980s. This unit is a reserve engine at Marine Corps Base Quantico, Virginia. CREDIT: Garry Kadzielawski

Naval Air Station Whiting Field's 1979 Dodge-Reading twinned agent unit built by Gibson was photographed in November 1989 at Whiting Field, Milton, Florida. CREDIT: Jim Atkinson

This yellow 1980 Fire-Tec Brush Structural rig was assigned to the Miramar Naval Air Station, San Diego, California, which later was consolidated within the Federal Fire Department, San Diego. CREDIT: Garry Kadzielawski

Mare Island Naval Shipyard, Vallejo, California, used this early 1980s Chevrolet with a Grumman step van body as a Quick Response Unit. CREDIT: Bill Killen Collection

This yellow 1980 Fire Trucks Inc.-Duplex with a 35-foot Tele-Squrt was assigned as Engine 61, San Diego Naval Shipyard, San Diego, California. This 1000-gpm pumper carried 500 gallons of water. CREDIT: Garry Kadzielawski

Philadelphia (Pennsylvania) Navy Shipyard's 1980 Spartan Motors-Pierreville 4-section 100-foot aerial ladder is one of five units delivered to the Navy. These units were powered with Cummins diesel engines and Allison MT-643 automatic transmissions and carried 193 feet of ground ladders. These single rear axle, closed canopy cab rigs were painted yellow. Other units were delivered to the Puget Sound Naval Shipyard, Bremerton, Washington; Naval Air Station, Alameda, California; Mare Island Naval Shipyard, Vallejo, California; and the Naval District of Washington. CREDIT: Glenn Vincent

This lime yellow 1980 GMC water tanker served as Naval Air Station Patuxent River's Crash 20 in Lexington Park, Maryland. CREDIT: U.S. Navy

Naval Submarine Base fire department, Groton, Connecticut's 1980 Fire Trucks Inc.-GMC Brush Structural pumper. The Navy purchased several combination Brush Structural pumpers. This unit was equipped with a 500-gpm pump and carried 600 gallons of water. CREDIT: Raoul K. Denton Collection

This 1980 Fire Tec-GMC Brush Structural rig was assigned to the Miramar Naval Air Station, San Diego, California. Designated "Brush 8," this yellow rig was equipped with a 750-gpm pump, and carried 400 gallons of water and 30 gallons of foam. CREDIT: Garry Kadzielawski

This 1980 Fire Trucks Inc.-Spartan served the Philadelphia Naval Shipyard, Philadelphia, Pennsylvania, as Engine 2. This yellow 1000-gpm pumper carried 500 gallons of water. CREDIT: Glenn Vincent

Three fire apparatus of the Naval Weapons Station, Yorktown, Virginia, are a 1979 Seagrave 750-gpm pumper, Oshkosh MB-1 Crash truck, and a Gibson water tanker. CREDIT: U.S. Navy

This 1981 Fire Trucks Inc.-GMC Brush Structural pumper was assigned to the Naval Air Station at Barber's Point, Hawaii. Equipped with a 500-gpm pump, this unit carried 600 gallons of water. CREDIT: Raoul K. Denton Collection

Naval Air Station Keflavik fire officers observe fire control operations from their Command Vehicle as firefighters extinguish fire in an aircraft that burst in to flames on impact at the Keflavik Naval Air Station, Keflavik, Iceland. CREDIT: U.S. Navy

This 1981 AM General 250-gpm brush truck carries 750 gallons of water and is assigned to the Naval Station Ingleside, Ingleside, Texas. CREDIT: Tom W. Shand

Naval Air Station Keflavik, Iceland, operated several 1962 FWD-1982 Quality Manufacturing P-2 Crash trucks. CREDIT: Bill Killen Collection

Naval District Washington firefighters man a 1982 Ward 79 1000-gpm pumper as they battle a multiple alarm fire at the Washington Navy Yard, Washington DC. CREDIT Bill Killen Collection

This lime yellow 1982 Ward 79 Ltd. was assigned to Naval Air Station North Island, Coronado, California. Notorious for problems with the cab and steering, these units were equipped with a 1000-gpm pump, and carried 750 gallons of water and 60 gallons of foam. CREDIT: Garry Kadzielawski

"Flowing Big Water" would be by Jim Manser's (former Fire Marshal Program Administrator) description of this Federal Fire Department San Diego training exercise with the Firefly III. Powered by a gas turbine engine, the Firefly III could deliver 5000 gpm from draft. CREDIT: U.S. Navy

This 1982 Emergency One, Inc. 110-foot 4-section rear mount aerial ladder is assigned to Marine Corps Base Camp Lejeune, North Carolina. CREDIT: Jim Atkinson

This 1983 Pierce-Chevrolet unit supported the Firefly I 2000-gpm gas turbine pump at the Federal Fire Department San Diego, California. Built for the U.S. Coast Guard, this was the original Firefly and was loaned to the Navy for test and evaluation. CREDIT: U.S. Navy

The pumping capability of Federal Fire Department San Diego's Firefly I was demonstrated across the United States, including the Nation's Capitol. CREDIT: James M. Manser, Retired Fire Marshal Administrator

The ability to rapidly deploy Marine Corps Crash crews and their fire apparatus is one of the unique requirements of the Marine Corps. Seen here is a Marine Corps P-19 crash truck being loaded on a C-130 aircraft during the first article acceptance tests. CREDIT: U.S. Navy

The primary purpose of this 1984 Chevrolet-Steeldraulics Products Inc. (SPI) support vehicle was to transport the monitor and provide a safe platform for flowing water supplied by the Firefly III pump. The U.S. Navy conducted several experiments and tests in delivering large volumes of water. The Firefly III was assigned to Naval Base Norfolk, Virginia, and then to Marine Corps Air Station Cherry Point, North Carolina. CREDIT: U.S. Navy

Jim Atkinson took this photograph of Naval Base Norfolk's 1966 American LaFrance Pioneer in 1984. The 750-gpm pumper was painted yellow and carried 500 gallons of water. Originally painted red, the Navy repainted most of their fire apparatus yellow in the early 1970s. CREDIT: Jim Atkinson Collection

This 1984 Walters-Duplex was the first structural pumper built by Walter Trucks, a long time manufacturer of airport firefighting vehicles. This lime green 1000-gpm unit carried 750 gallons of water and 100 gallons of foam and served as Engine 9 at Naval Air Station Miramar, San Diego, California. CREDIT: Garry Kadzielawski

This 1984 Seagrave Model HR-07DF is one of two 4-section 100-foot aerial ladders delivered to the Navy in 1984. This unit, USN 7400058, designated "Truck 2" was assigned to Naval Air Station North Island, Coronado, California. These lime green units had H model Seagrave cabs, single rear axles, and carried 163 feet of ground ladders. CREDIT: Garry Kadzielawski

Jim Atkinson photographed this 1984 Chevrolet twinned agent unit with a Reading utility body at Naval Air Station New Orleans in 1989. The lime green rig has a reflective stripe lengthwise in the middle of the body. Credit: Jim Atkinson

Washington, DC, fire department assists Naval District Washington at a multiple alarm blaze in the Washington Navy Yard, Washington, DC, in 1984. CREDIT: Bill Killen Collection

Naval Base Norfolk's 1984 Chevrolet-Steeldraulics Products Inc. (SPI) and the trailer mounted Firefly III gas turbine powered pump. The 5000-gpm monitor mounted on the Chevy was manufactured by FEECON. The gas turbine aircraft engine drove the Austrian manufactured Ochsner pump. It carried a large assortment of LDH adapters, Akron Portable Hydrants, special Akron Apollo monitors with 5-inch intakes, and 5-inch and 3-inch hose. Various tips can be seen mounted on the top of the compartments. CREDIT: U.S. Navy

The Children's Fire Safety House made its debut at the International Association of Fire Chiefs conference in New Orleans in 1985. The brainchild of former Pensacola Naval Air Station Fire Chief Sid Booze, the Children's Fire Safety House has been credited with saving the lives of several children as a result of the training provided at elementary schools in northwestern Florida and southern Alabama. CREDIT: NAS Pensacola Fire Department

This factory photograph was taken during acceptance tests conducted by the Navy of the 1985 Peter Pirsch 110-foot senior aerial in Kenosha, Wisconsin. Shown here painted and lettered to exact Navy specifications, the truck passed all tests and was accepted by the Navy for assignment to Naval Station Treasure Island, San Francisco, California. This Pirsch Model CSN6 was powered with a 6-cylinder Detroit Diesel engine and Allison Automatic transmission. CREDIT: Bill Killen Collection

THE PIRSCH "SKYTOP 110" CUSTOM SENIOR AERIAL

This Pirsch advertisement depicts the Naval Station Treasure Island's 110-foot aerial AFTER the Pirsch factory dressed it up with a "Sunfire" finish paint job and Hand Gold Leaf Lettering. The 110-foot all-aluminum heavy duty 4-section ladder included a telescoping waterway, and high strength aluminum ladder. CREDIT: Bill Killen Collection

This 1985 Ward 79 Ltd. all-wheel drive Duplex chassis was specifically designed for the Navy's South Pole facility, Williams Field, McMurdo Station, Antarctica. This unit was powered with a Caterpillar 3406T diesel engine, equipped with a 750-gpm pump, and carried 3500 gallons of water. CREDIT: Raoul K. Denton Collection

Firefighters at Naval Air Station Willow Grove, Horsham, Pennsylvania, designed and installed a protective "apron" to prevent damage and chipping of the paint when personnel donned self-contained breathing apparatus mounted in the high side compartment. CREDIT: Bill Killen

This 1985 GMC Tanker equipped with a 500-gpm pump and a 1500-gallon water tank was assigned to Guantanamo Bay Naval Station, Guantanamo Bay, Cuba. CREDIT: Jim Derstine

Naval Communications Station Cheltenham, Maryland, received this 1986 Pierce Dash 1000-gpm pumper in 1986. Pierce built 61 of these units for the Navy, three for the Veterans Administration and one for the U.S. Coast Guard. These two-door, cab over, tilt cab units were powered by Cummins L-10 300-hp diesel engines with Allison HT-740 automatic transmissions. The units were equipped with a Waterous 1000-gpm pump, a 750-gallon water tank, and a 100-gallon foam tank. The Akron Brass mounted above the pump compartment delivered over 1000 gpm during acceptance tests and is controlled at the pump panel. This was the inaugural of the Navy's new paint scheme of white over lime green with an 8-inch reflective stripe above the wheel line on both sides of the body. CREDIT: Bill Killen Collection

Federal Fire Department San Diego Fire Chief Don Crutchfield, now retired, obtained this box style trailer from the U.S. Customs office San Diego, California. Fire department personnel, using design concepts from the Maryland Fire & Rescue Institute, University of Maryland, constructed this self-contained breathing apparatus/maze mobile training laboratory for training firefighters. CREDIT: U.S. Navy

Norfolk Naval Base Firefly III Team operating at draft from a stream on Marine Corps Base Camp Lejeune, North Carolina. The Firefly III delivered over 2000-gpm fire flow via 5-inch hose over a distance of one mile to supply an aerial water tower, and two interior firefighting operations using four 1 1/2-inch hose lines. CREDIT: U.S. Navy

This military 1-ton 4 x 4 1986 GMC 1000 Series was equipped with a 200-gpm pump and a 250-gallon water tank for brush firefighting operations. CREDIT: John Rieth

This 1986 Morita-Nissan diesel powered 750-gpm pumper is assigned to Marine Corps Base Camp Butler, Okinawa, Japan. Designated Reserve Engine 1, this unit was remanufactured in 1998. CREDIT: Stuart Cook

Naval Weapons Station Earle fire department, Colts Neck, New Jersey, uses this 1986 GMC tractor to pull a 1990 Etnyre 4000-gallon water tanker equipped with a 250-gpm pump. CREDIT: John Rieth

This "before" picture shows Charleston Naval Station's 1974 American LaFrance 1000 Series truck in 1985 after a new 85-foot 3-section ladder was installed. The body and paint were in rough shape as seen here in this picture. CREDIT: Bill Killen Collection

This "after" picture shows the results of the "self-help" initiative of firefighters at the Charleston Naval Station, Charleston, South Carolina. Captain Michael Stallings supervised and coordinated the repairs and painting. The white over lime green rig carried 208 feet of ground ladders. Charleston Naval Station closed in 1997 and the truck was sent to the Defense Reutilization Management Office for disposal. CREDIT: Bill Killen Collection

Naval Air Station Patuxent River fire department used this 1980s vintage Boston Whaler 25-foot Frontier as a fire rescue boat for operations in the Patuxent River and the Chesapeake Bay. The boat was assigned to the Solomons Island station and served a dual function for security patrol as well as fire and rescue operations. The boat was powered by twin 150-hp outboard Johnson motors and used a standard Navy 250-gpm portable pump and supplied a prepiped monitor mounted on the bow. CREDIT: U.S, Navy

Fleet Combat Training Center Dam Neck fire department in Virginia Beach, Virginia, used this 1980s vintage Dodge as Brush 52. The unit carried a slide-in unit with a 250-gpm pump and 200 gallons of water. CREDIT: Bill Killen Collection

Marine Corps Base Quantico, Virginia's 1987 Pierce Arrow 55-foot Tele-Squrt. This white over red rig is designated as Squad 3 and is equipped with a 1000-gpm pump and carries 500 gallons of water and 1000 gallons of Aqueous Film Forming Foam. CREDIT: Bill Killen Collection

This white over red 1987 Pierce Arrow served as Engine 4, Marine Corps Base Quantico, Virginia. The truck has a 1000-gpm pump, and carries 750 gallons of water and 100 gallons of foam. CREDIT: Bill Killen Collection

This 1987 Pierce Arrow 55-foot Tele-Squrt pumper is assigned to the Naval Surface Warfare Center, Indian Head, Maryland. The white over lime green rig is equipped with a 1000-gpm midship mounted pump, a 500-gallon water tank, and a 100-gallon foam tank. CREDIT: Bill Killen Collection

Naval Air Station Lakehurst, Lakehurst, New Jersey, firefighters earned recognition for modifying a 1967 International water tanker in 1987. One of their innovative changes included the 1990s look on the front of the truck. CREDIT: Bill Killen Collection

The Navy's Mobile Training Laboratory–HAZMAT is believed to be the first of its kind ever developed for the training of emergency services personnel in hazardous materials incident command and control, hazardous materials identification, and leak and spill control. This mobile training unit was designed and built by Naval Air Station Jacksonville, Florida, fire department personnel in 1986. CREDIT: U.S. Navy

Interior view of the Command and Control section of Naval Air Station Jacksonville's Hazardous Materials Mobile Training Laboratory photographed in Baltimore at Firehouse Expo 1987. CREDIT: Maryland Fire & Rescue Institute

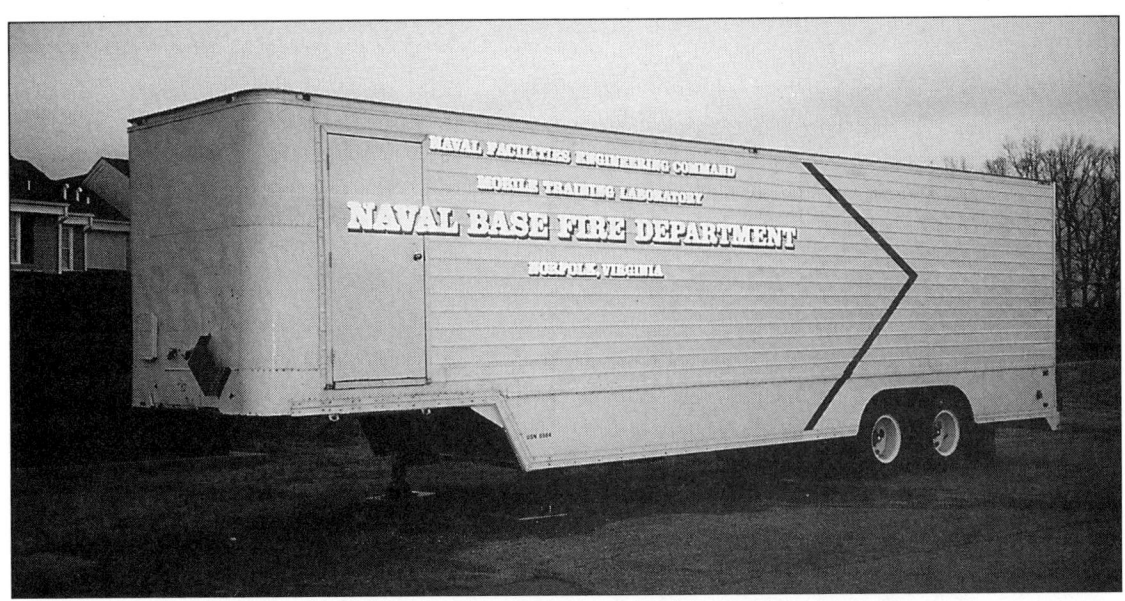

Norfolk Naval Base Norfolk, Virginia, firefighters constructed this Mobile Training Laboratory (MTL) as a self-help project funded by the Naval Facilities Engineering Command. The MTL is designed to simulate conditions found in structures and is used to train firefighters to work in confined spaces wearing self-contained breathing apparatus. This is one of 10 MTLs built by Navy firefighters and has been used extensively in training municipal and volunteer firefighters in Virginia as well as Navy firefighters in the Tidewater, Virginia, region. CREDIT: Bill Killen

Oceana Naval Air Station fire department's (Virginia Beach, Virginia) 1988 Pierce Arrow 1000-gpm pumper operating at a structural fire in the Wherry Housing, NAS Oceana. CREDIT: Warren Dixon

These Navy and Marine Corps 1988 Pierce Arrow pumpers are ready for delivery as seen here at Pierce Manufacturing's Appleton, Wisconsin, plant. CREDIT: Bill Killen

The Naval Station fire department's fleet of structural fire apparatus at Guantanamo Bay, Cuba, circa February 1988. CREDIT: Jim Derstine

This 1988 Pierce Arrow 100-gpm pumper was assigned to the Naval Facility at Argentia, Newfoundland, Canada. When the Navy closed the Argentia facility, the truck was turned over to the Canadian government. CREDIT: Paul Rakowski

Naval Station Long Beach Fire Department's white over lime green 1988 Pierce Arrow 1000-gpm pumper carried 750 gallons of water, 100 gallons of Aqueous Film Forming Foam, and was one of 61 units delivered to the Navy. Ten units painted white over red were delivered to the Marine Corps. These were powered by Cummins L-10 300-hp diesel engines with Allison Automatic HT-740 transmissions. These rigs had two-door Arrow Cabs, triple crosslays above the pump and left side high compartments. CREDIT: Tom W. Shand

1988 Pierce Arrow 50-foot Tele-Squrt 1000-gpm pumper served Naval Surface Warfare Center Dahlgren, Virginia, and Naval Surface Warfare Center, White Oak, Maryland, and now serves the Naval Ship Systems Engineering Support Station, Philadelphia, Pennsylvania. CREDIT: NSWC Dahlgren Fire Department

Norfolk Naval Base fire department obtained this early 1960s LARC from surplus and equipped it with portable fire pump and rescue equipment. This unit is used for fire suppression and rescue operations at the Norfolk Naval Base Complex, Norfolk, Virginia. CREDIT: Jim Atkinson Collection

This 1988 Chevrolet crew cab is powered with a 6.2L diesel engine and equipped with a Slagle slide in firefighting package for brush firefighting. The tank carries 275-gallons of water with a booster reel mounted on the water tank and a 250-gpm pump. CREDIT: U.S. Navy

This 1988 Morita pumper is assigned to the Consolidated Fleet Activities Fire Department Sasebo, Japan. CREDIT: U.S. Navy

This 1989 Pierce Arrow 105-foot rear mount aerial ladder, assigned to the Long Beach Naval Station, Long Beach, California, is one of three units delivered to the Navy. Painted white over lime green, each unit was powered by a Detroit Diesel 8V-92TA 475-hp diesel engine with an Allison HT-740 4-speed automatic transmission. Smeal manufactured the 4-section steel ladder and mounted them to the Pierce chassis at the Smeal plant. These four-door Arrow cab, tandem axle units carried 163 feet of ground ladders. Two other Pierce Arrow aerial ladders were assigned to Naval Air Station Corpus Christi, Texas, and Naval Air Station Jacksonville, Florida. CREDIT: Tom W. Shand.

The Navy's Service Life Extension Program funded the rehab of Pensacola Naval Air Station's 1979 Fire Trucks Inc.-Spartan 750-gpm pumper. The Construction Battalion Center Gulfport, Mississippi, completed the rebuild in 1989. CREDIT: Jim Atkinson

This 1989 Pierce Arrow is assigned to the Norfolk Naval Base, Norfolk, Virginia, and is one of two tractor-drawn aerials in first line service in the Navy. This two-door Arrow canopy cab is painted white over lime green, carries a complement of 163 feet of ground ladders, and is powered by a Detroit Diesel 8V-92TA, 475-hp diesel engine with an Allison HT-740 automatic 5-speed transmission. The aerial ladder is a 105-foot 4-section steel ladder. CREDIT: Tom W. Shand

Norfolk Navy Shipyard Fire Department personnel mounted this 1989 3/4-ton Chevrolet flat bed with a hose bed equipped with two 1 1/2-inch hoses. This unit was specifically designed for responding to the Norfolk Naval Hospital Parking Garage in Portsmouth, Virginia. The chassis was replaced in 1996 with a Dodge Ram 3/4-ton chassis. CREDIT: U.S. Navy

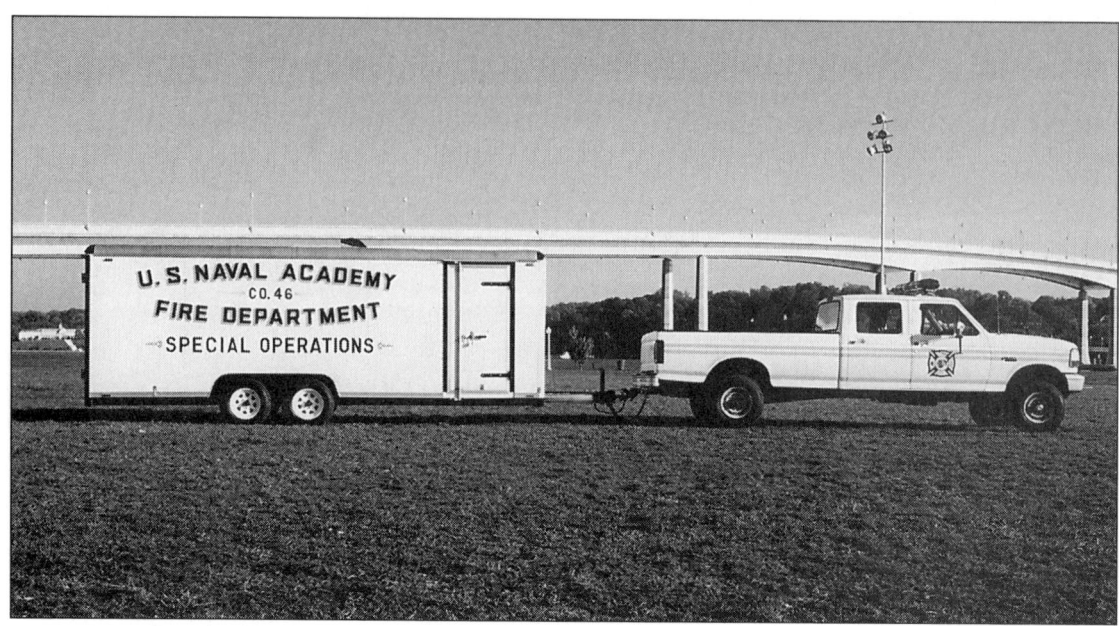

The U.S. Naval Academy's Special Operations trailer and 1989 Ford Crew Cab pickup support emergency operation at the Naval Academy and the David Taylor Research Station in Annapolis, Maryland. CREDIT: Jerry Sack

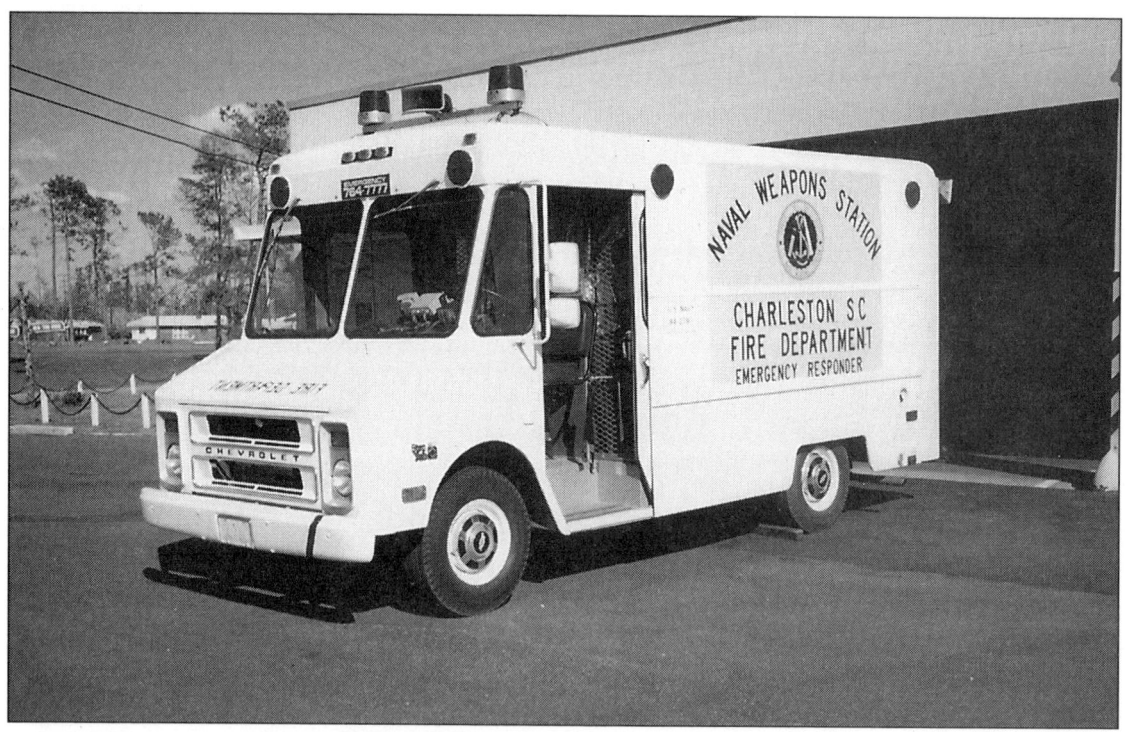

Naval Weapons Station Charleston, South Carolina's 1989 Chevrolet Step Van is used as a support vehicle for emergency operations. CREDIT: Bill Killen

Naval Security Group Edzell Scotland used British manufactured fire apparatus to protect the Navy facility at Edzell. The year and manufacturer of these trucks is unknown. CREDIT: U.S. Navy

Engine 911 was built by Naval Air Station Pensacola firefighters and is a major component of the fire department's public fire safety education program along with the Children's Fire Safety House seen in the background. CREDIT: NAS Pensacola Fire Department

Imagination, some sheet metal, and a golf cart were transformed into Naval Air Station Pensacola's Engine 911. The little engine actually pumps water and is a hit with the kids during Fire Prevention Week activities. CREDIT: NAS Pensacola

Ward 79 Ltd. built this 3000-gallon tanker on a tandem axle R Model Mack for Naval Weapons Station Earle, Colts Neck, New Jersey. The rig is powered by a Mack Maxidine 350-hp engine and an Allison Automatic HD740 4-speed transmission and has a Hale fire pump. CREDIT: Tom W. Shand

This unique one-of-a-kind rescue combination is a 1990 International Water Rescue unit with a boom and a rubber boat. CREDIT: John Rieth

This 1990 Pierce Dash 4 x 4 Rescue was assigned to the Marine Corps Air Station El Toro, Orange County, California, until it was closed in 1997. CREDIT: Richard Adelman Collection

This 1990 Pierce Arrow 65 Tele-Squrt powered by L10 Cummins engine with a 4-speed Allison Automatic transmission was delivered to the Naval Surface Warfare Center, Dahlgren, Virginia. This unit is equipped with a Waterous 1000-gpm pump, 500-gallon booster tank, 100-gallon foam tank, carries 800 feet of 5-inch large diameter hose, 400 feet of 3-inch supply hose, and medical equipment including an Automatic External Defibrillator. CREDIT: Bill Killen

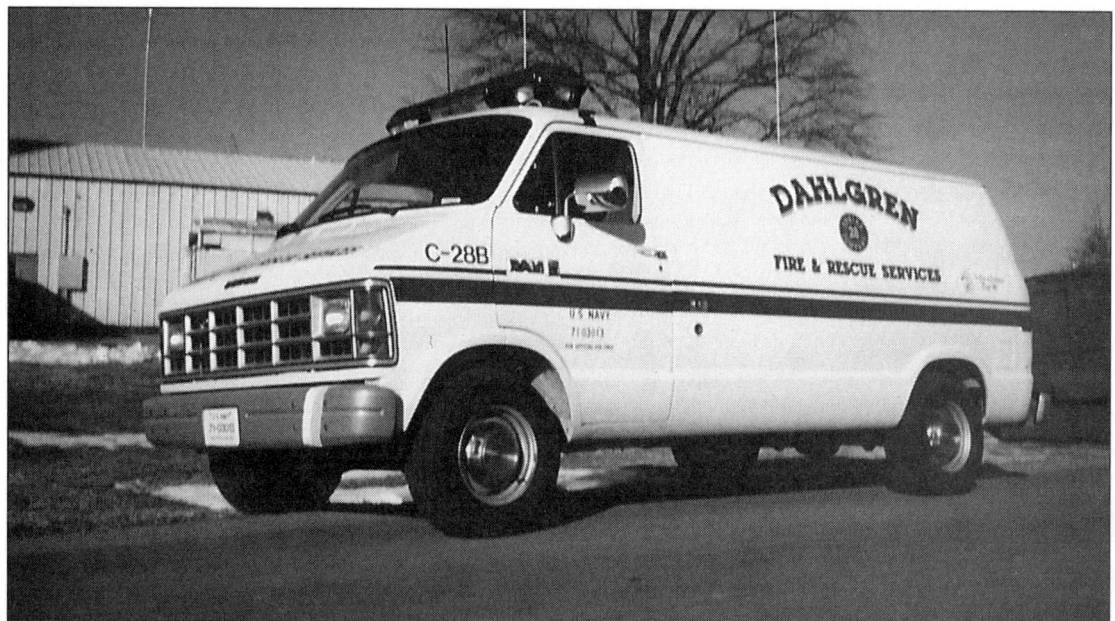

Naval Surface Warfare Center, Dahlgren, Virginia, converted this 1990 Dodge 1/2-ton van into a Command van. This vehicle is the primary response vehicle for the Assistant Fire Chief for each shift. This unit contains a Toshiba laptop computer and a Hewlett Packard printer, various books and references, including maps for utilities, as well as a complete set of prefire plans for every building on the Naval base. CREDIT: Bill Killen

This 1990 Grumman Firecat 1000-gpm pumper is assigned to the North Atlantic Treaty Organization (NATO) Fuel Facility at Ponta Delgado, San Miguel Island, Azores, Portugal. CREDIT: John J. Wentzel

Naval Inventory Control Point's Engine 237 is a 1990 Duplex chassis with a 1992 Fire Apparatus Unlimited body. Engine 237 is one of five units which was part of a Ward 79 Navy contract for 18 1250-gpm pumpers. Ward defaulted and the Navy took possession of five Duplex D-350 two-door chassis. Fire Apparatus Unlimited completed the construction and delivered two units to Naval Support Facility Thurmont, and one to the Naval Academy. One unit went to the Veterans Administration. CREDIT: Bill Killen

This Pierce Arrow 65-foot Snorkel Tele-Squrt pumper is assigned to the U.S. Naval Academy, Annapolis, Maryland, and is listed in the Navy inventory as a 1990 Pierce Arrow. CREDIT: Bill Killen Collection

Consolidated Fleet Activities Fire Department Sasebo, Japan, operates this 1990 Morita pumper at the Naval Station Sasebo, Japan. CREDIT: U.S. Navy

This 1991 Chevrolet Rescue unit with air cascade system assigned to the Puget Sound Naval Shipyard now serves the Navy Regional Fire Department Northwest's Kitsap Peninsula area in the State of Washington. CREDIT: Lynn Johnson

Lakehurst Naval Air Engineering Center Fire Department's 1991 Humvee brush rig is equipped with a 150-gpm fire pump and a 200-gallon water tank. CREDIT: Dennis C. Sharpe

This 1991 Pierce Dash 4 x 4 is equipped with a 1000-gpm pump, 750-gallon water tank, and 100-gallon foam tank. Several of these units are assigned to the Navy Fleet Hospital. CREDIT: Tom W. Shand

Naval Surface Warfare Center Dahlgren Division fire department's Hazardous Materials Response trailer is equipped to mitigate hazardous materials spills and leaks on the installation. This Navy unit provides emergency response to hazardous materials incidents off the naval station into King George County, Virginia, as part of a mutual aid agreement. CREDIT: Bill Killen

Naval Ship Parts Control Center Mechanicsburg's 1991 Pierce Arrow 65-foot Tele-Squrt sports a new enclosed four-door cab, air conditioning, and new front and rear axles with anti-lock brakes as a result of a November 1999 rehab. CREDIT: Bill Killen

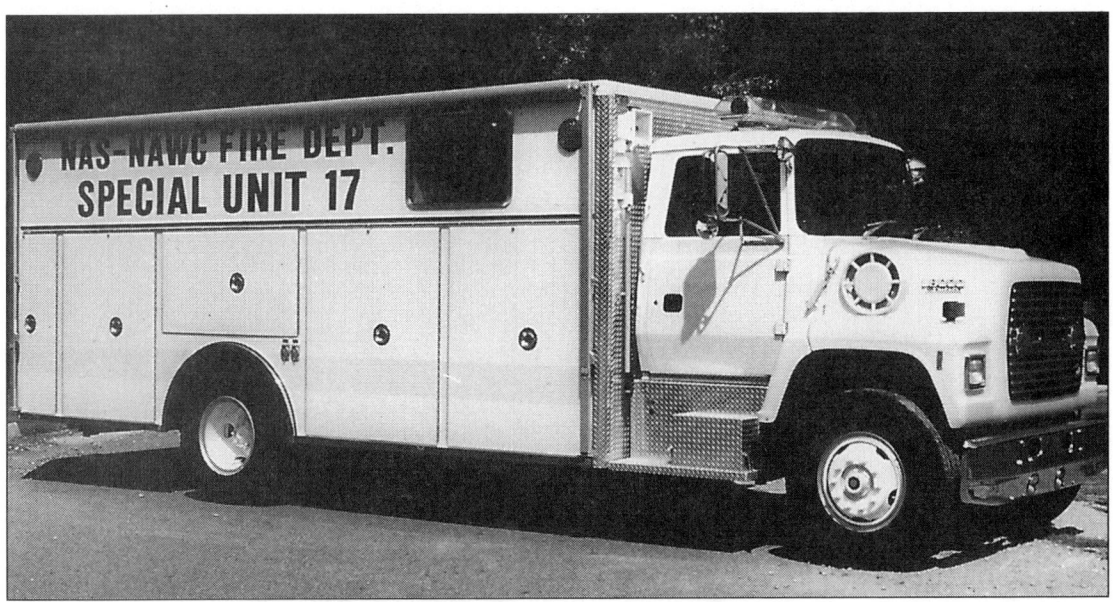

This 1991 Pierce Hazardous Materials Emergency Response vehicle, assigned to the Naval Air Station Patuxent River, Lexington Park, Maryland, is one of 26 units built on a Ford LN 8000 chassis and powered by a Ford 240-hp diesel engine driven by an Allison MT 643 4-speed automatic transmission. CREDIT: Bill Killen Collection

This 1991 Pierce Arrow 65-foot Snorkel Tele-Squrt 1000-gpm pumper is assigned to Naval Station Ingleside, Ingleside, Texas, and is one of seventeen units delivered to the Navy. These white over lime green Pierce Arrow two-door cab units included high side compartments and triple cross lays. The power plant is a Cummins L-10 320-hp diesel engine with an Allison HT 740 automatic transmission and a Waterous 1000-gpm centrifugal pump. These rigs carried 500 gallons of water and 100 gallons of foam. CREDIT: Tom W. Shand

This 1991 Simon-Duplex D350/1994 Fire Apparatus Unlimited is one of five units built for the Navy after Ward 79 Ltd. defaulted in 1992. The U.S. Naval Academy's Engine 461 is equipped with a 1000-gpm pump, 750-gallon water tank, and 100-gallon foam tank. CREDIT: Joe MacDonald Collection

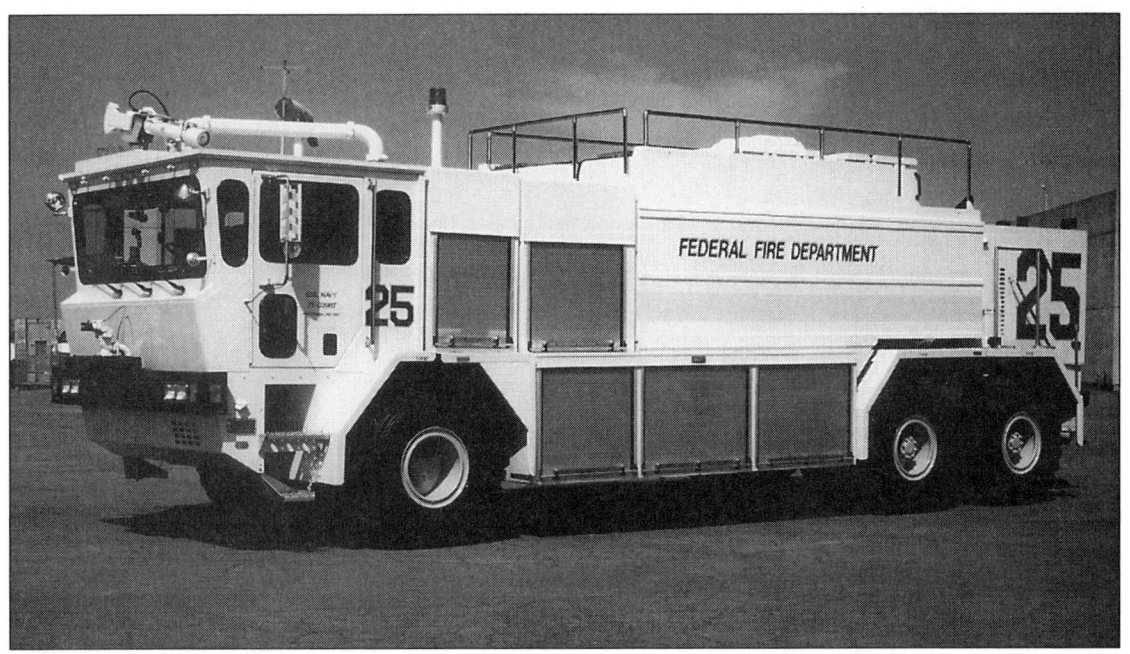

Naval Air Station Fallon Nevada's 1992 Oshkosh TA-3000 crash truck is powered by a Detroit Diesel 8V 92TA diesel engine and an Allison HT 750DR 5-speed automatic transmission. The single-stage centrifugal 2000-gpm supplies a non-aspirating type roof turret (600/1200 gpm) and bumper turret (300 gpm). The turrets are electric joystick control with auto-oscillation and infinitely variable pattern from straight stream to fully dispersed. CREDIT: Tom W. Shand

This 1986 Morita 4 x 2 pumper is built on an Isuzu chassis and is assigned to Camp Foster, Marine Corps Base Okinawa, Japan. The pump is a two-stage 750-gpm pump and the 2000-liter water tank is approximately 375 U.S. gallons. CREDIT: T. Kevin King

Marine Corps firefighters conduct training evolution at Camp Butler, Okinawa, Japan, with their 1986 Morita 35 meter ladder. This steel 5-section rear mounted ladder is equipped with an escape sled and a two-stage 750-gpm pump. This ladder was built on a Nissan chassis with a four-door five-man cab and painted red with a white cab. CREDIT: T. Kevin King

Rescue 2 was manufactured by the Morita Company on a 1987 Nissan 4 x 4 chassis. This red and white rig is equipped with a 750-gpm pump and carries 2000 liters of water. CREDIT: T. Kevin King

Marine Corps Base Camp Pendleton California's 1992 Pierce Arrow 100-foot rear mount aerial is equipped with a 1000-gpm pump. The white over red truck carries 300-gallons of water and 30-gallons of foam. CREDIT: Richard Adelman Collection

Norfolk Naval Shipyard's 1993 Pierce Arrow 105-foot 3-section aerial tower is equipped with a 1500-gpm pump and a complement of 163-feet of ground ladders. Powered by a 475-hp Detroit Diesel 8V-92TA and an Allison automatic HT-740 transmission. The four-door Arrow Cab and tandem axle unit is painted white over lime green. CREDIT: Tom W. Shand

Fleet Industrial Supply Center Craney Island's 1993 Pierce Arrow 50-foot Tele-Squrt is equipped with a 1250-gpm midship Waterous pump, a 1400-gpm FEECON around the pump foam proportioning system, and carries 500 gallons of water and 100 gallons of foam. The water tank was designed to be used as a foam tank if desired and has two 50-gallon foam tanks. CREDIT: Bill Killen

1993 was the last year of manufacture of the Amertek 1000-gallon airfield rescue firefighting vehicle. This lime green Amertek is Naval Air Station Fallon, Nevada's unit number 27. CREDIT: Tom W. Shand

Navy's southernmost fire department was McMurdo Station, Antarctica, where this 1984 Ward-Duplex 3500-gallon tanker with a 1000-gpm pump provided a mobile water supply. The 1993 Pierce Arrow 1000-gpm pumper carried 750 gallons of water and 100 gallons of foam. CREDIT: Tom W. Shand Collection

This 1994 Pierce Arrow 65-foot Tele-Squrt is assigned to Naval Air Station Fallon, Nevada. Equipped with a 1250-gpm pump, this rig carries 500 gallons of water and 100 gallons of foam. CREDIT: Tom W. Shand

The U.S. Navy Staten Island Homeport operated this 1994 Chevrolet HAZMAT truck for the short period of time the Naval base was operational. This truck was transferred to the city of New York fire department. CREDIT: John M. Calderone

Three 1994 Pierce Arrow pumpers for MCB Quantico, Virginia; Marine Corps Air Ground Combat Center, Twentynine Palms, California; and Marine Corps Logistics Base, Barstow, California, at a cost of $185,000.00 per vehicle. Each unit includes 750-gallon tank, 1250-gpm Waterous pump, Detroit Diesel 6V091 engine, Allison HT-740 automatic transmission, and FEECON around-the-pump foam system. These were the first four-door enclosed pumpers in the Marine Corps inventory. A total of 5 units were purchased between 1994 and 1996. CREDIT: T. Kevin King

This 1993 Pierce Arrow 105-foot, 4-section rear mount aerial ladder is Naval Air Station Pensacola's Ladder Company 1 and serves Naval Air Station Pensacola, Corry Station and Saufley Field, Pensacola, Florida. CREDIT: NAS Pensacola Fire Department

Naval Air Engineering Center Lakehurst's 1994 Pierce Dash is a white over lime green 1250-gpm pumper and carries 750 gallons of water and 100 gallons of foam. This four-door enclosed cab rig has a top mounted pump panel and is nearly identical to their 1996 Pierce Dash pumper. The only visible difference is the Federal Siren mounted on the front bumper of the 1994 model. CREDIT: Dennis C. Sharpe

This 1994 GMC 4 x 4 pickup serves as the Naval Air Engineering Center Lakehurst, New Jersey, Fire Chief's car and Incident Command vehicle. The unit is equipped with a fiberglass cap and slide-out Command Module. CREDIT: Dennis C. Sharpe

Norfolk Naval Base Fire Department's Engine 6 is a 1994 KME Renegade 1250-gpm pumper. This white over lime green rig features a four-door enclosed cab, high side compartments with roll up doors and a prepiped monitor mounted above the pump compartment. Designated a Type I, the Navy version carries 750 gallons of water and 100 gallons of foam. CREDIT: Tom W. Shand

THE KME ADVANTAGE

American-Engineered, Custom-Built Fire Apparatus

One Industrial Complex Nesquehoning, PA 18240 717-669-5132

KME Renegade 1250 GPM Custom Pumper

Type 1 As Built For The U.S. Navy

This KME Renegade 1250-gpm Custom Pumper is one of four types built under the first standardized Department of Defense structural pumper specification during 1994 and 1995. Features include four-door enclosed cabs and high side compartments with roll up doors. The Navy units are white over lime green and have smaller foam tanks. This Type 1 was delivered to Naval Training Center Great Lakes, Chicago, Illinois. CREDIT: KME Fire Apparatus

Brush Structural pumper is assigned to Federal Fire Department San Diego. This unit is equipped with a Hale 250-gpm pump and carries 500 gallons of water. This unit was delivered in March 1994 at a cost of $91,000.00. CREDIT: Jaimie Wood

Naval Weapons Station Earle firefighters assist the Colts Neck, New Jersey, volunteer fire department at a structural fire in March 1995. CREDIT: John Rieth Collection

Philadelphia Naval Shipyard's Marine 1 is a former U.S. Coast Guard 32-foot rescue boat with a 500-gpm pump and served the Navy complex in Philadelphia, Pennsylvania. CREDIT: John M. Calderone

Four, 1995 AM General Hummer chassis wildland firefighting vehicles built for Marine Corps Base Camp Pendleton, California, by Fire Attacker Fire Apparatus, Petersburg, Michigan, at a cost of $70,237.00 each. Each vehicle has a 300-gallon tank, Robwen Class A foam system, and a Darley 1-1/2 AGE fire pump. CREDIT: T. Kevin King

Naval Weapons Station Earle firefighter operates a master stream from the Pierce Tele-Squrt pumper during a mutual aid call in Colts Neck, New Jersey. CREDIT: John Rieth Collection

In honor of Fire Prevention Week and the Naval Academy's 150th Anniversary, a ribbon cutting ceremony took place October 11, 1995 to dedicate the U.S. Naval Academy's new 1995 KME 1250-gpm fire engine. Academy Superintendent Admiral Charles R. Larson received the keys to the new fire engine from Fire Chief Jerry Sack and "Sparky the fire dog." CREDIT: U.S. Naval Academy

Naval Weapons Station Earle's 1995 Ford L Series 4 x 4 tractor and 1990 Etnyre 4000-gallon water tanker. CREDIT: John Rieth

Fleet Industrial Supply Center Craney Island Fire Department's mission is to protect the largest fuel farm in the Navy, located in Portsmouth, Virginia. This 1995 UPF (United Plastic Fabricating) Foam Supply trailer carries 1000 gallons of foam and is equipped with a 5-hp Briggs and Stratton powered transfer pump. CREDIT: Bill Killen

The Naval Station Pascagoula, Mississippi, Fire Chief uses this 1996 Dodge Ram pickup as a command vehicle. CREDIT: Bill Killen

This 1996 Kovatch 102-foot Aerial Tower is assigned to Naval Air Station Patuxent River, Lexington Park, Maryland, and was an "add-on" order from the U.S. Air Force aerial tower contract. This lime green unit differs from the Air Force version in that it has a KME Excel four-door cab with partial raised roof, whereas the Air Force model is a cab-forward low profile four-door cab. A Detroit Series 60 diesel engine and an Allison HD-4060 automatic transmission provide power. This tandem axle rig has a 1500-gpm midship pump, carries 150 gallons of water and a complement of 115 feet of ground ladders. CREDIT: Bill Killen Collection

Naval Air Engineering Center Lakehurst's Rescue 3 was formerly a utility truck on a 1996 Ford F5000 chassis. Fire department personnel built this rescue unit as a self-help project for use as a support vehicle for airfield operations. CREDIT: Dennis Sharpe

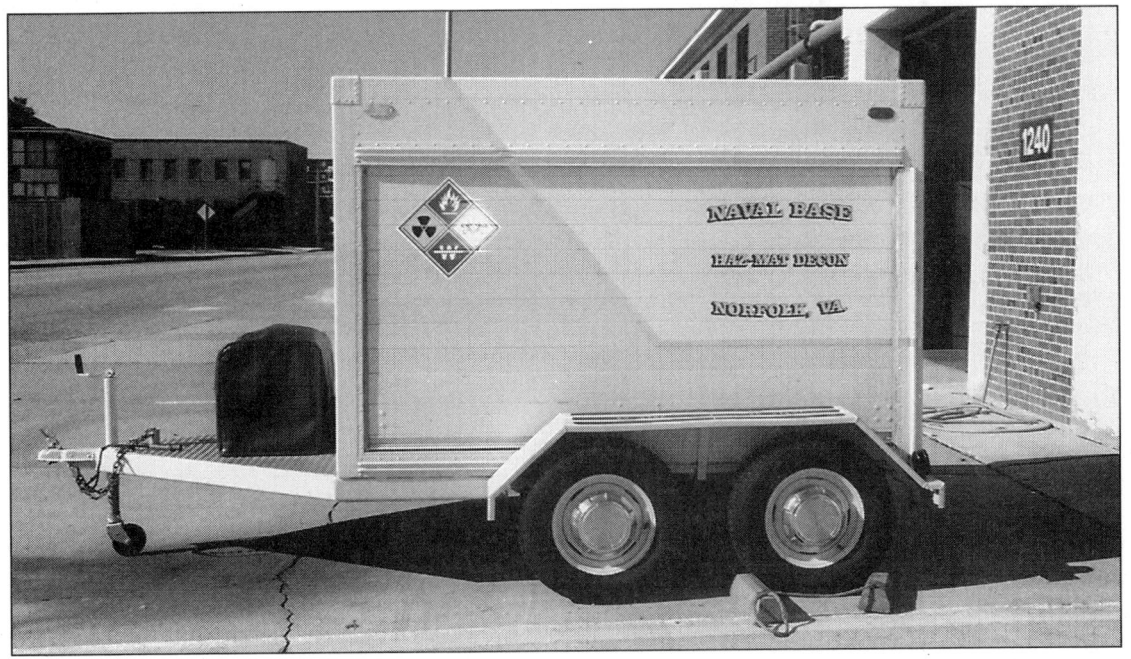

Norfolk Naval Base fire department personnel built this HAZMAT Decon Trailer from a surplus trailer and excess equipment obtained from the salvage yard. CREDIT: Chief Danny Miller

Naval Air Engineering Center Lakehurst, New Jersey's 1996 Ford 4 x 4 Crew Cab Rescue. The white over lime green rig also tows a Knapheide trailer with additional equipment. CREDIT: Dennis C. Sharpe

This 1996 Pierce Arrow 1250-gpm pumper is equipped with a 500-gallon water tank and a 50-gallon foam tank and serves the Naval Air Engineering Center Lakehurst, New Jersey. CREDIT: Dennis C. Sharpe

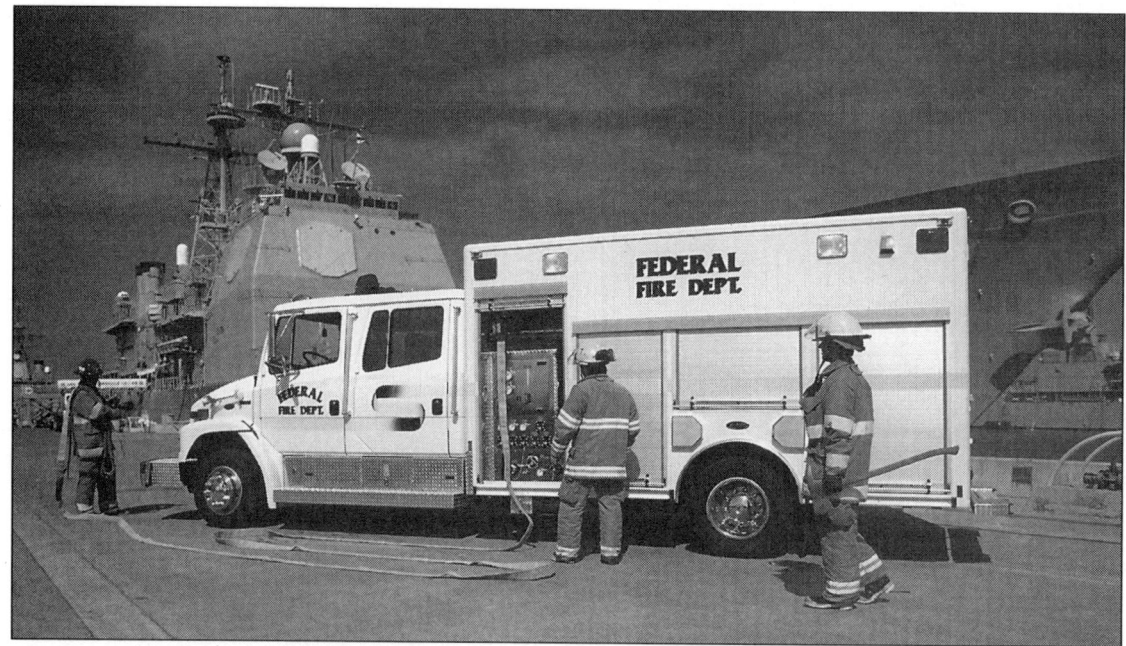

Emergency One's 1996 Trident pumper ambulance assigned to the Navy Federal Fire Department in San Diego, California. This multiple mission vehicle is a fully KKK (d) (ambulance specifications) certified rescue transport with fire suppression and rescue capabilities, and was delivered in March 1996 at a cost of $125,000. CREDIT: Emergency One

This 1997 Pierce Arrow 105-foot 4-section steel rear mount aerial ladder carries a complement of 163 feet of ground ladders and is assigned to the Consolidated Fleet Activities Fire Department, Yokosuka, Japan. CREDIT: U.S. Navy

Hackney Corporation manufactured Naval Station Norfolk Fire Department's 1997 Technical Rescue unit. The fire pump, tanks, and body were removed from a 1986 Pierce Dash and replaced with a Hackney rescue body in August 1997. CREDIT: Chief Danny Miller

Naval Base Norfolk's Fire Prevention engine is a big hit with children, especially when the Fire Prevention Clown is at the wheel. CREDIT: Fire Inspector John Hallman III

Marine Corps Base Camp Butler's Engine 12 is a 1986 Nissan Diesel 750-gpm pumper manufactured by Morita in Japan and remanufactured in 1997. CREDIT: Stuart Cook

This 1997 Oshkosh T-1000 is assigned to Naval Air Station Norfolk, Hampton Roads Regional Fire Department. The T-1000 is slightly longer than the Navy Oshkosh P-19 and has the same Detroit Diesel V-92TA diesel engine used in the Oshkosh TA-3000 trucks. This white over lime green rig carries 1000 gallons of water and 130 gallons of foam. CREDIT: Tom W. Shand

This 1997 Pierce Arrow 4-section steel 105-foot four rear mount aerial ladder is assigned to the Portsmouth Naval Shipyard, Portsmouth, New Hampshire. This lime green and white rig is powered by a Detroit Diesel with an Allison HT-740 automatic transmission and carries 163 feet of ground ladders. CREDIT: Dick Bartlett

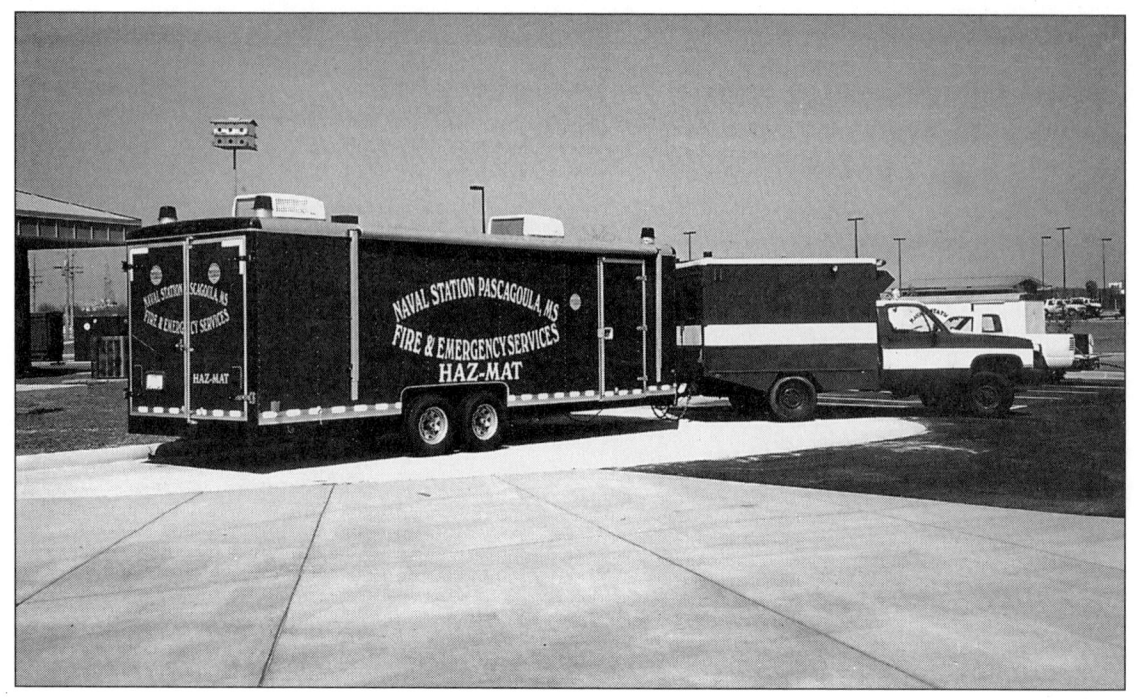

Naval Station Pascagoula's 1997 Wells Cargo HAZMAT emergency response trailer is towed by 1984 Chevrolet M1010 that formerly served Naval Air Station Cecil Field, Florida. CREDIT: Bill Killen

This 1978 Seagrave Rear Admiral 100-foot aerial ladder was retrofitted with a KME cab by the Public Works Center Pensacola and shipped to the Naval Station Newport in 1997. CREDIT: Tom Gemp

Federal Fire Department San Diego operates this Light & Air Trailer pulled by a 1980 International Harvester tractor. This unit is equipped with a 12 CFRM air compressor, 3 SCBA bottle refilling stations, a Bobcat with a bucket scoop and extensive equipment and supplies to control and mitigate hazardous materials spill and leak emergencies. CREDIT: Jamie Wood

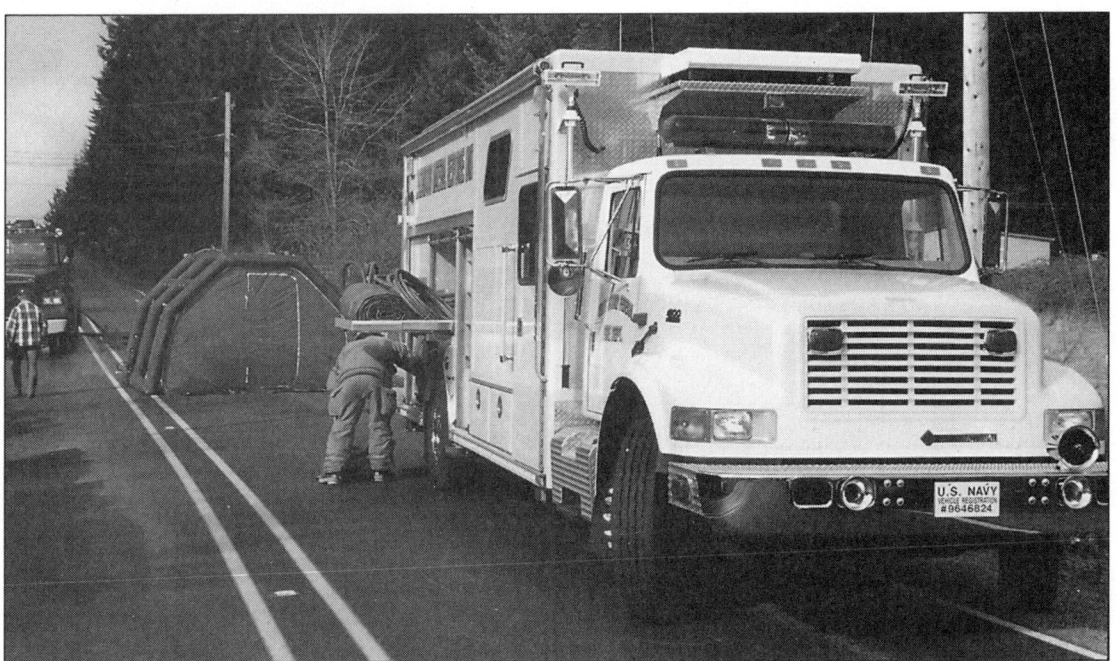

Puget Sound Naval Shipyard's 1998 Emergency One HAZMAT Command, on an International Chassis costing $185,000.00, is seen here providing mutual aid at a methamphetamine lab house fire. CREDIT: Kitsap County Fire District 7

Puget Sound Naval Shipyard's 1998 E-One ladder tower and engine in mutual aid at a structure fire in downtown Bremerton, Washington. CREDIT: Lynn Johnson

This 1998 Pierce Saber 1250-gpm pumper carries 750 gallons of water and 100 gallons of foam and is assigned to Naval Surface Warfare Center Indian Head, Maryland. CREDIT: Bill Killen

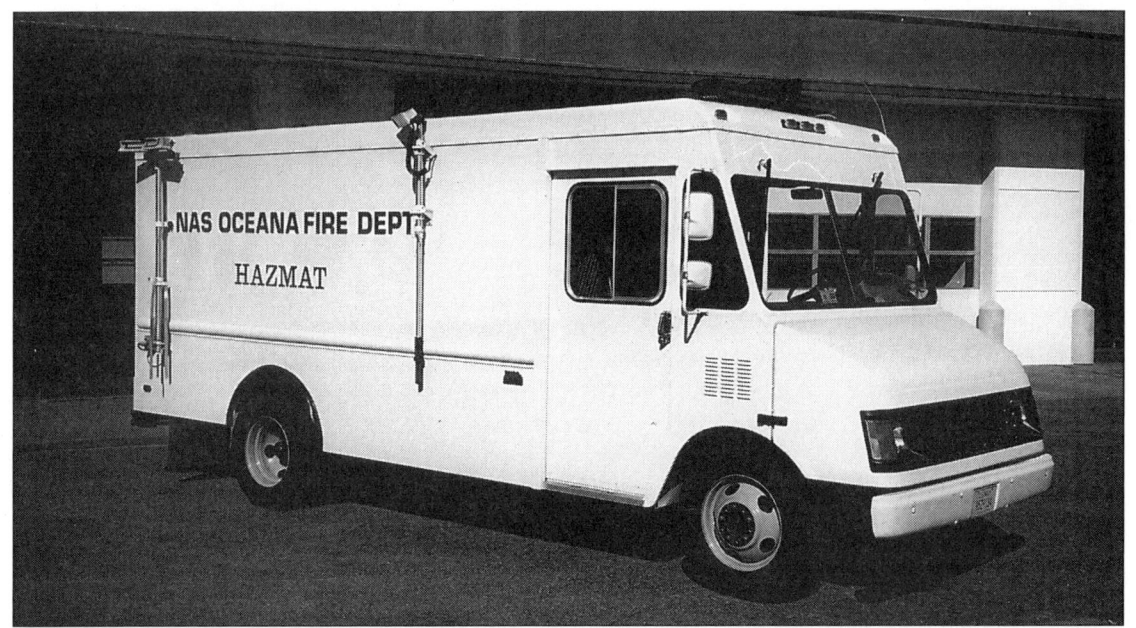

The Environmental Office at Naval Air Station Oceana provided this completely equipped HAZMAT van to the Oceana Naval Air Station fire department for response to hazardous materials spills and leaks. CREDIT: Bill Killen

Marine Corps Air Station Miramar fire department's 1998 Brush truck was built by Master Body Works, Incorporated of Anaheim, California. Built according to California Division of Forestry specifications, the truck has a wheelbase of 152 inches, which allows access to all fire trails in the western states. The hydrostatic drive to a Waterous pump permits pump and roll operations and at the same time maintains constant pressure regardless of rpm. The truck has a 500-gallon water tank and a 20-gallon Class A foam tank. The approach angle is 26 degrees and the departure angle is 22 degrees. The unit was delivered in July 1999 at a cost of $150,000.00. CREDIT: Stuart Cook

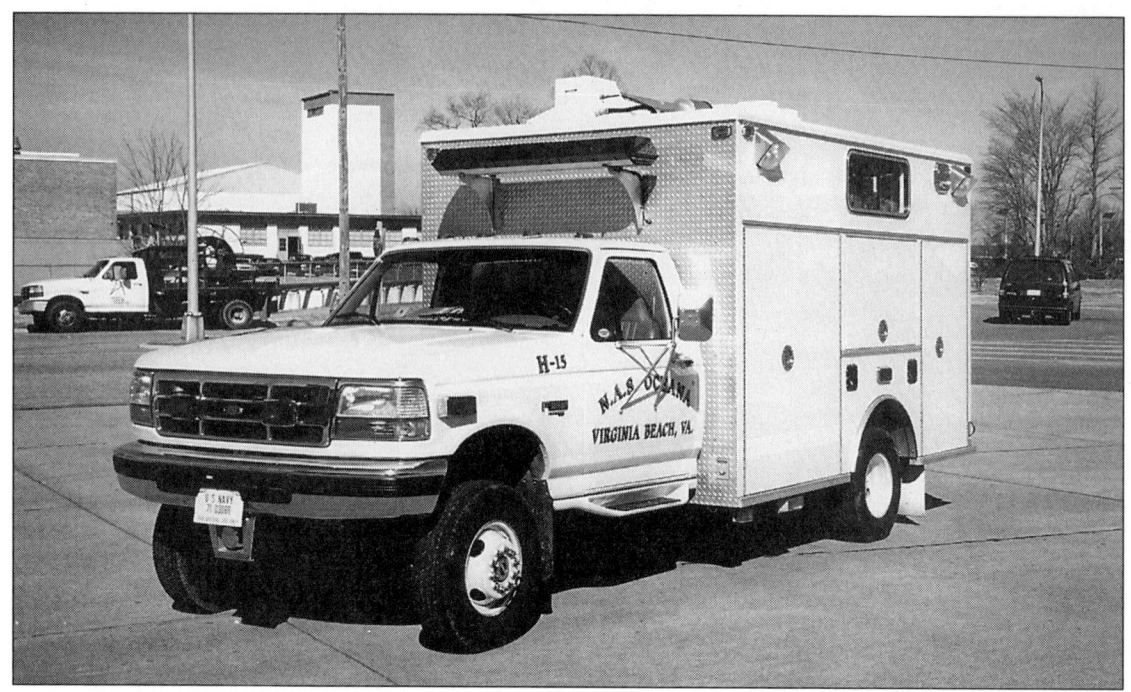

Naval Air Station Oceana's KME Rescue is built on a 1998 Ford F Series Super Duty chassis. This unit provides rescue support to the airfield and other emergency operations. CREDIT: Bill Killen

This 1998 Emergency One Cyclone TC with a mid-mount 50-foot Teleboom is assigned to Naval Air Station Oceana, (Virginia Beach) Navy Regional Fire Rescue Hampton Roads, Virginia. Powered with a 350E Cummins diesel with Allison transmission and Rockwell anti-lock brakes. The unit has a Hale 1250-gpm pump, a 500-gallon water tank, and 100 gallons of 3% AFFF. E-One delivered this white/lime green four door enclosed cab unit in August 1998 at a cost of $302,000.00. CREDIT: Bill Killen

Puget Sound Naval Shipyard firefighters provide decontamination in mutual aid to Kitsap County Fire District 7 at a methamphetamine lab house fire in 1999 in Kitsap County, Washington. CREDIT: Kitsap County Fire District 7

This 1999 Pierce Dash 2000 with a full rescue body, 1250-gpm pump, 750-gallon polytank with 100 gallons of foam serves Naval Surface Warfare Center Dahlgren, Virginia. The unit is equipped with hydraulic ladder rack, carries 1000 feet of 5-inch large diameter hose, 300 feet of 3-inch supply hose, and 2,250 feet of 1 3/4-inch crosslays. Extrication equipment includes airbags, air tools, and Hurst extrication tools. Medical equipment includes an oxygen kit and an Automatic External Defibrillator. CREDIT: Bill Killen

District Chiefs with the Pensacola Naval Air Station, Pensacola, Florida, are assigned 1999 Dodge Dakota's for command vehicles. These vehicles are leased from the Public Works Center, Pensacola, Florida. CREDIT: Pensacola Naval Air Station Fire Department

This 1999 Oshkosh T-1500 is one of the latest acquisitions in the Navy fire apparatus fleet, assigned to Naval Air Engineering Center Lakehurst, New Jersey. The roof of this lime green rig is painted white and features both roof and bumper turrets. The water and foam tank capacities are 1500 and 220 gallons respectively. CREDIT: Dennis C. Sharpe

U.S. Naval Academy Fire Department's 1999 Ford Expedition Command vehicle has mutual aid radio frequencies for the City of Annapolis and Anne Arundel County. CREDIT: Joe MacDonald Collection

This 1999 Emergency One D150 Foam Tanker, the first vehicle to be lettered "Navy Regional Fire Rescue Hampton Roads," is assigned to the Fleet Industrial Supply Center Fuel Depot Craney Island, Portsmouth, Virginia. Designated "Foam 409," the unit is built on an International chassis and powered with a Navistar DT-466 diesel engine. It has a 150-gpm Edwards bronze transfer pump, carries 1500 gallons of 3% AFFF foam, and cost $125,000.00. CREDIT: Bill Killen

This 1983 FMC Corp. Fire Trac tracked vehicle and 1980 AM General tractor and trailer were assigned to the Naval Air Station Cecil Field for brush firefighting operations. The unit was transferred to Naval Station Mayport when Naval Air Station Cecil Field closed in October 1999. CREDIT: U.S. Navy

Identified as "Federal Fire Department Engine 1," this unit is one of two 1999 Emergency One Cyclone Rescue pumpers delivered to the Naval Inventory Control Point, Mechanicsburg, Pennsylvania. These white over lime green 1250-gpm pumpers carry 750 gallons of water and 40 gallons of Class A and B foams. Acquisition cost was $213,000.00 each. CREDIT: Bill Killen

Driver Operator Roy Bullock operates Naval Weapons Station Earle's Engine 94-90 at a working fire early morning on October 31, 1999. The 1988 Pierce Arrow 1000-gpm pumper is supplying three handlines and a master stream. CREDIT: Monmouth County Fire Investigator Chris Pujat

"Whiting 108" is a 2000 Ford F-350 Super Duty 4 x 4 Crew Cab powered by 7.3-liter Ford Power Stroke Diesel and is assigned to OLF Holley Field approximately 20 miles south of Naval Air Station Whiting Field in Milton, Florida. "Whiting 108" has an automatic transmission, a 5000-pound Warren winch mounted on the front bumper, and carries 200 gallons of Aqueous Film Forming Foam, 200 pounds of Halon, and 550 cubic feet of Nitrogen. "Whiting 108" replaced a 1985 Chevrolet twinned agent unit. CREDIT: Chris Hatch

Navy and Marine Corps Bases

About the Author

The author is currently the Director, Navy Fire and Emergency Services and began his fire service career at the age of 16 as a volunteer with the Potomac Heights Volunteer Fire Department, Indian Head, Maryland. In June 1960, he was appointed a Probationary Firefighter with the Naval Ordnance Station Fire Department, Indian Head, Maryland.

His paid fire service career spans forty years and includes service with the Kennedy Space Center Fire Department where he served on the Apollo and Skylab Astronaut Rescue Teams, Fire Chief of the Lake Barton Fire Control District, Orange County, Florida; Senior Faculty member, Maryland Fire and Rescue Institute, University of Maryland, Fire Chief, Metropolitan Washington Airports, and was appointed to his current position in 1985.

The author represents the federal government on the Commission on Fire Accreditation International, serves as the Vice-President of the International Association of Fire Chiefs Foundation and is an active member of the International Association of Fire Chiefs. He has authored papers for fire service technical publications and co-authored the history of the Kennedy Space Center Fire Department.

He is an avid fire apparatus history buff and restored a 1923 Stoughton fire engine and is currently writing the history of Stoughton Wagon Works. He is a member of the Society for the Preservation and Appreciation of Motorized Fire Apparatus in America, the Antique Truck Club of America, the Antique Automobile Club of America, and the Antique Truck Historical Society.

He is a Past Master of Collington Lodge #230 Ancient Free & Accepted Masons, Bowie, Maryland and regularly presents lectures on George Washington. He and his wife Carole have four children and six grandchildren.

MORE GREAT BOOKS

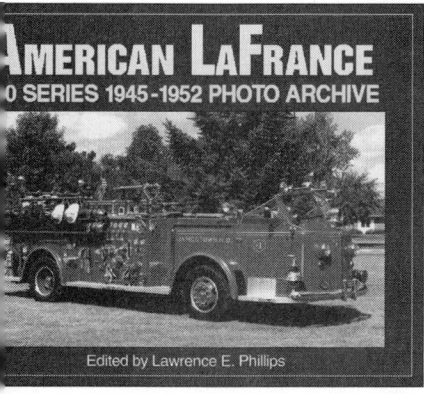

AMERICAN LAFRANCE 700 SERIES 1945-1952 PHOTO ARCHIVE
ISBN 1-882256-90-5

AMERICAN LAFRANCE 700 & 800 SERIES 1953-1958 PHOTO ARCHIVE
ISBN 1-882256-91-3

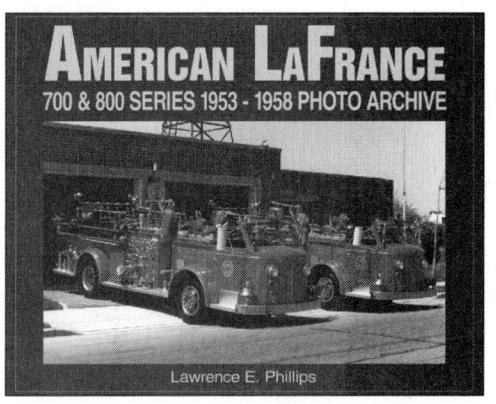

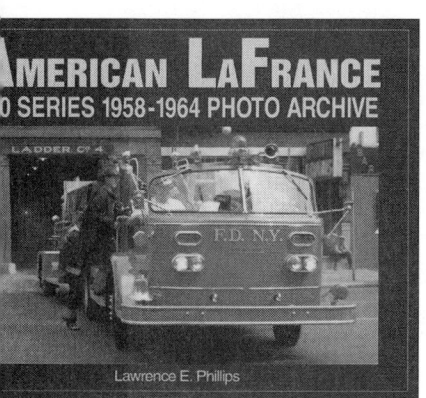

AMERICAN LAFRANCE 900 SERIES 1958-1964 PHOTO ARCHIVE
ISBN 1-58388-002-X

AMERICAN LAFRANCE 700 SERIES 1945-1952 PHOTO ARCHIVE VOL. 2
ISBN 1-58388-025-9

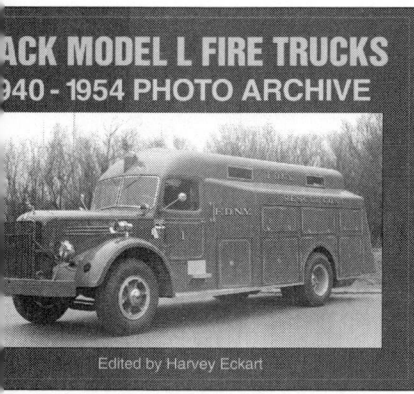

MACK MODEL L FIRE TRUCKS 1940-1954 PHOTO ARCHIVE*
ISBN 1-882256-86-7

MACK MODEL C FIRE TRUCKS 1957-1967 PHOTO ARCHIVE*
ISBN 1-58388-014-3

MACK MODEL B FIRE TRUCKS 1954-1966 PHOTO ARCHIVE*
ISBN 1-882256-62-X

MACK MODEL CF FIRE TRUCKS 1967-1981 PHOTO ARCHIVE*
ISBN 1-882256-63-8

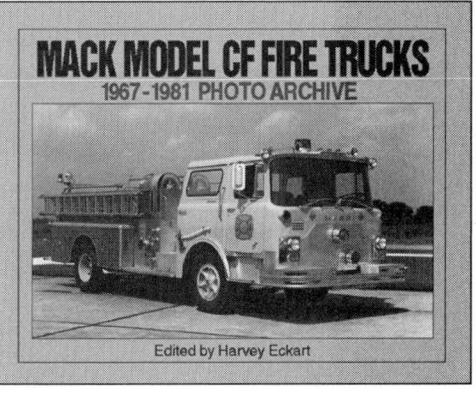

All books available through:
Iconografix, Inc. PO Box 446/BK, Hudson, Wisconsin, 54016
Telephone: (715) 381-9755, (USA) (800) 289-3504, Fax: (715) 381-9756

MORE GREAT BOOKS

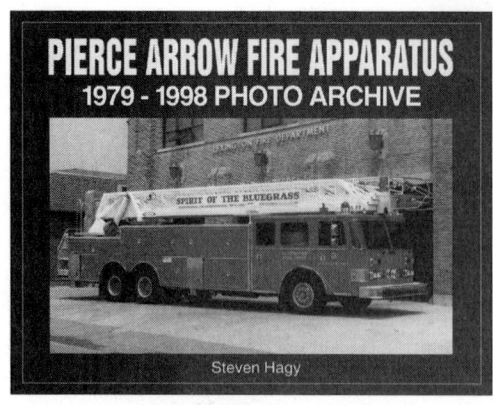

PIERCE ARROW FIRE APPA-
RATUS 1979-1998 PHOTO
ARCHIVE
ISBN 1-58388-023-2

LOS ANGELES CITY FIRE
APPARATUS 1953 - 1999
PHOTO ARCHIVE
ISBN 1-58388-012-7

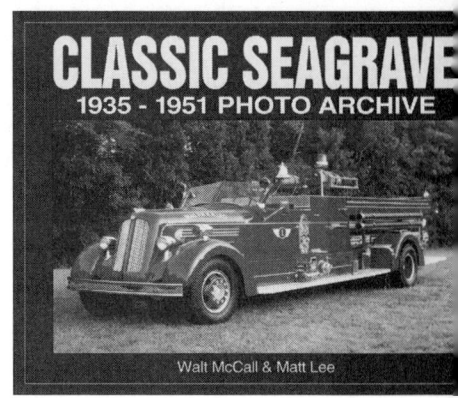

YOUNG FIRE EQUIPMENT
1932-1991 PHOTO
ARCHIVE
ISBN 1-58388-015-1

CLASSIC SEAGRAVE 1935-
1951 PHOTO ARCHIVE
ISBN 1-58388-034-8

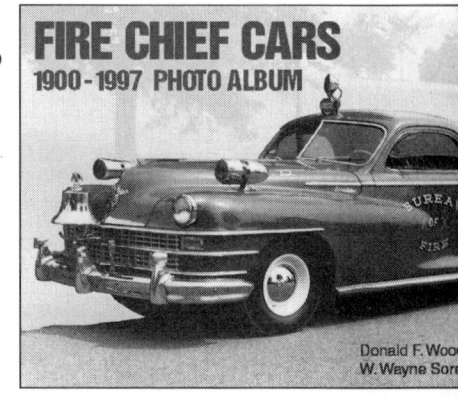

WARD LAFRANCE FIRE
TRUCKS 1918-1978 PHOTO
ARCHIVE
ISBN 1-58388-013-5

FIRE CHIEF CARS 1900-
1997 PHOTO ALBUM
ISBN 1-882256-87-5

SEAGRAVE 70TH ANNIVER-
SARY SERIES PHOTO
ARCHIVE
ISBN 1-58388-001-1

VOLUNTEER & RURAL
FIRE APPARATUS PHOTO
GALLERY
ISBN 1-58388-005-4

All books available through:
Iconografix, Inc. PO Box 446/BK, Hudson, Wisconsin, 54016
Telephone: (715) 381-9755, (USA) (800) 289-3504,
Fax: (715) 381-9756